Learning Computer Programming

It's Not About Languages

Learning Computer Programming

It's Not About Languages

Mary E. Farrell

Charles River Media, Inc.
Hingham, Massachusetts

Publisher: David Pallai
Production: Publishers' Design & Production Services
Cover Design: The Printed Image

CHARLES RIVER MEDIA, INC.
20 Downer Avenue, Suite 3
Hingham, Massachusetts 02043
781-740-0400
781-740-8816 (FAX)
info@charlesriver.com
www.charlesriver.com

This book is printed on acid-free paper.

Mary Farrell. *Learning Computer Programming: It's Not About Languages.*
ISBN: 1-58450-061-1

Library of Congress Cataloging-in-Publication Data

Farrell, Mary.
Learning computer programming / Mary Farrell.
p. cm.
ISBN 1-58450-061-1 (paperback : acid-free paper)
1. Computer programming. I. Title.
QA76.6 .F379 2002
005.1—dc21

2002000259

Printed in the United States of America
02 7 6 5 4 3 2 First Edition

Requests for replacement of a defective CD-ROM must be accompanied by the original disc, your mailing address, telephone number, date of purchase, and purchase price. Please state the nature of the problem, and send the information to CHARLES RIVER MEDIA, INC., 20 Downer Avenue, Suite 3, Hingham, Massachusetts 02043. CRM's sole obligation to the purchaser is to replace the disc, based on defective materials or faulty workmanship, but not on the operation or functionality of the product.

For my parents.

Contents

Preface

Welcome to CyberRookies *Learning Computer Programming: It's Not About Languages.*

This book is an introduction to the topic of computer programming.

The focus of the book is on the common elements of programming in any language. As you might know, every couple of years, a new language comes along that is the most popular language and *the one* to learn! Java is that language now. Yet, a programmer who is flexible and confident should realize that that is the nature of programming. There will always be some new hot language around the corner and the pressure will be on to learn that one too.

What I hope to accomplish in this book is to introduce the beginner to the common threads in any programming language. Once you understand how to program loops and decisions, you can then learn how to work with interesting data structures like arrays and records. Then you can move on to interesting programming concepts like recursion, searching, and sorting data. With an elementary background such as the one provided in this book, you can approach each new language with the objective: what makes this new language special and what do I need to learn to master it?

Every new language is designed with some particular advantage over already existing languages. When you learn a new language, the easiest thing to do is to discover the way the new language handles your familiar topics that you know from other languages. Once you master those, the trick is to master the new features offered by the language.

The first few chapters are dedicated to discussing such elements as learning how to store information in holders - called variables - and later manipulating that information. You will learn how to write assignment statements, how to show information (display output) on the screen, and the importance of the algorithm in developing a solution to a problem. After deciding on the best way to solve a problem, you can then start to program.

In Chapters 5 and 6 you learn how to allow the computer to make a decision and how to construct a loop. A loop allows the programmer to have the computer execute some task repeatedly.

In Chapters 7 and 8 you learn about using functions in programming - that is, developing a function, or separate block of code, to solve a problem. Once a function has been written, you can call on it whenever you need it instead of having to rewrite that code. Chapter 8 discusses the topic of graphics in programming and employs functions to draw various shapes on the screen.

The later chapters focus on data structures that will be with you in almost any language: e.g. the array and the record. Chapter 11 is about the use of external files. Advanced topics such as pointers, searching, sorting, and recursion are introduced in Chapters 12 through 15. The last three chapters of the book are dedicated to 3 specific languages: HTML, C++, and Java.

Chapter 16 focuses on HTML since it is so useful for the Internet. Although it is not as complex a language as other high level languages, its usefulness cannot be overestimated.

Chapter 17 is all about C++ and its powerful capabilities, centering around the introduction of classes and objects. Object oriented programming presented a major shift in programming development, and C++ is one of the most powerful languages to use this concept.

Chapter 18 is about Java and why it is so useful and prevalent today. Java is fully object oriented and came at the right time. The explosion of the World Wide Web and Java is one of the happiest coincidences in technology in the last decade.

Acknowledgments

I would like to thank both my father and mother for their multiple readings of this manuscript. I would also like to thank the following people for their help in answering my questions: Lauren Crocker, Charlie Drane, Alessandro Costa, Joel Lamousnery, and Amy Cole. I would like to thank Dave Pallai for his great patience and Matt Pallai for introducing me to his father. I would like to thank all my students, past and present. Their questions made me a better teacher and a better learner.

CHAPTER

1

First Things First

IN THIS CHAPTER

- Important Hardware Components and Software
- The Concept of Digitization
- Binary Digits, Bits, Bytes
- The Computer, an Electronic Machine
- Languages: High Level vs. Low level
- A First Look at the Algorithm and Its Relationship to Programming

1.1 INTRODUCTION

Computer technology is all around us—represented not only by personal computers but also by ATM machines, fuel injection in cars, digital cameras, and telephone communications. None of this technology would be possible without computer programming. Programming provides instructions that tell the machine how to operate. Everything the personal computer does—from playing video games and music to typing simple documents in Microsoft *Word*—requires a set of instructions.

This book will provide you with a basic knowledge of computer programming. It will tell you how to control the computer so that it can repeatedly perform a given function, make a decision, and store information in the right kind of holder—contingent on what that information looks like. These are just a few of the things you will learn.

It is important to cover certain fundamentals before learning to program. The more you understand the machine and how it works, the easier it will be for you to grasp the concepts of programming.

1.2 HARDWARE AND SOFTWARE

A computer is usually hooked up to a printer, an external drive, and various other peripheral devices such as scanners, modems, and the like. All the physical components of a computer make up its *hardware*. Think of the word "hardware" as describing those parts of a computer that are "hard" or can be touched. You can touch a printer but you cannot touch programs that are running on a computer. Programs represent *software*. All programs—whether they are commercial programs, games, CD-ROMs, word processing applications, or programs that make the computer itself operate—are examples of software.

Programs are sets of steps that tell a computer what to do. There are directions for everything that happens in the computer. For example, saving a file on the hard drive occurs because, inside the computer, a program is "telling" the computer to save a file rather than delete it. Saving, printing, and deleting a file are just a few of the programs called *system programs.* They are also referred to as the *operating system.*These programs are the programs that allow the computer to handle its basic operations: opening and closing, saving, and deleting files. Likewise, getting an application such as Microsoft *Word* to open up and start running is the task of the operating system.

Application programs are programs sold on the market today (e.g., *Doom,* Microsoft *Word,* Adobe *Premier,* Norton *Utilities,* etc.). The word "application" refers to a set of programs "applied" to some real life task to make it easier to do. Some of the earliest application programs—Microsoft *Word, WordPerfect, ClarisWorks*—were created to facilitate typing long documents.

HINT!

System software is the set of programs that enables the computer to store and retrieve data, save and delete files, and, operate application programs.

Let's look at a list of the main parts of a computer:

Keyboard
Mouse
Printer
Hard Drive
External Drive
RAM
CPU

THE KEYBOARD, MOUSE, AND PRINTER

The *keyboard* and the *mouse* are used to communicate with the computer. The keyboard is used for typed commands whereas the mouse is used for "clicking" and interacting with the graphical user interface (G. U. I.). The printer delivers on paper what is on the screen.

HARD DRIVE VS. EXTERNAL DRIVE

The *hard drive* is the internal memory of a computer. Application programs are saved there, as are the system files. Think of the hard drive as the amount of permanent storage space that a computer has. A house with a basement and an attic has much more storage area than an apartment with only closets for storage. Computers with large drives (e.g., 2 gigabytes) are in demand because of their capacity to save large applications on their hard drives. Metrowerks *Code Warrior*™ for example, needs 80 MB of storage space. The game *NBA Basketball 2000*™ needs 75 MB of storage.

An *external drive* is a drive used to expand storage capacity outside of the computer. As recently as the early 1990s, many personal computers had little memory in the hard drive, necessitating storage outside of the computer. Memory needed for application programs at that time was not what it is today.

RAM: RANDOM ACCESS MEMORY

The *RAM* is really the workspace for the computer. Think of the RAM as a large table on which you place many different things. If you have a small workspace, you can't open too many things at once because they won't fit on the table. If your operating system uses 10 MB, while Microsoft *Word* uses 4 MB and *Doom II* uses 4 MB, you could have both applications open at the same time as the operating system (total of 18), if your computer has 32 MB of RAM, for example. Now, if you run a game such as *NBA Basketball 2000™*, you'll need 32 MB just for that application alone! Running the operating system with that game brings your total to 42 MB and that's why you'll want a computer with a good size RAM (e.g., 64 MB) if you usually play such games. Opening several applications at once and not running out of memory is desirable.

RAM is temporary memory that is lost once the computer is turned off. The term *volatile* is used to describe this memory because anything not saved will be lost in a sudden abrupt fashion (e.g., if a plug is suddenly pulled on a machine or you have to reset your computer for some reason, everything in the RAM will be lost).

ROM: Read Only Memory

ROM or *read only memory,* is memory that is not lost when the machine is turned off. Certain instructions for the computer are permanently etched onto a chip at the time the computer is manufactured. Thus ROM is permanent memory.

CENTRAL PROCESSING UNIT

The main part of the computer is the *Central Processing Unit* or *C.P.U.* This is where the computer stores, processes, and retrieves data. The C.P.U. manages all the functions of the computer including processing data—manipulating data by sending it from one place to another—or by performing some math on the data. The C.P.U. contains the *arithmetic/logic unit, A.L.U.,* and the *control* unit of a computer.

THE A.L.U.: ARITHMETIC/LOGIC UNIT

The *A.L.U.* is where programmers are most affected. When you write a program for a computer, its A.L.U. will be called into use to perform some math (the arithmetic part) or to evaluate a decision (the logic part) by the programmer. That's why we need the arithmetic/logic unit. The logic portion is the part of the unit that can handle decisions. Most interesting programs need to have the ability to make a decision.

THE CONTROL UNIT

The *control unit* of the C.P.U. is used to regulate program flow.

This unit executes statements in sequence and will only repeat steps or skip steps if programmed to do so.

If you are writing a program and put a number such as "5" into a holder called "x" and decide to increase the value of "x" from "5" to "7," the A.L.U. will be used to do this task. If you want to print the message, "I have had enough!" 250 times on a computer screen, the control unit will have work to do, because the programmer is asking the computer to do something repeatedly. Thus the programmer controls how long or how many times the computer does something; in this case, printing a message 250 times.

It was mentioned earlier that the computer can store, process, and retrieve data, and the components described in this section handle that work.

1.3 USING NUMBERS TO REPRESENT INFORMATION: DIGITIZATION

To understand how the computer does what it does, we can examine a situation in the real world to see how numbers can be used to give information about that situation.

Imagine four towns, roughly equidistant, as shown in the drawing. A bad snow storm has closed some of the roads between these towns whereas others remain open.

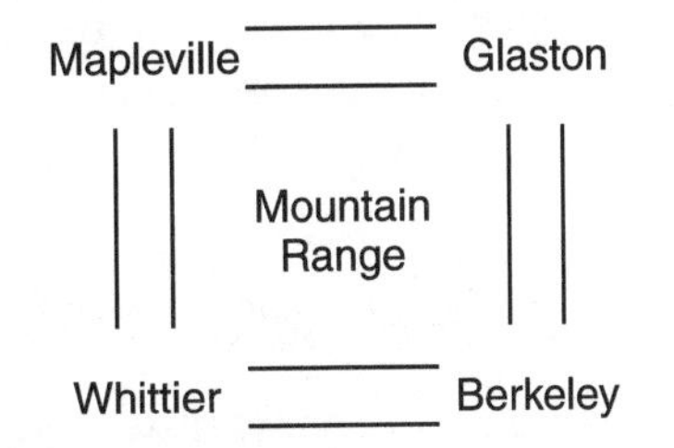

Between Mapleville and Glaston the roads are closed as are the roads between Whittier and Mapleville. The roads between Glaston and Berkeley are open and the roads between Berkeley and Whittier are open. If we let the number one (1) represent the roads being open and the number zero (0) represent the roads being closed, then we can redraw the picture to look like the following:

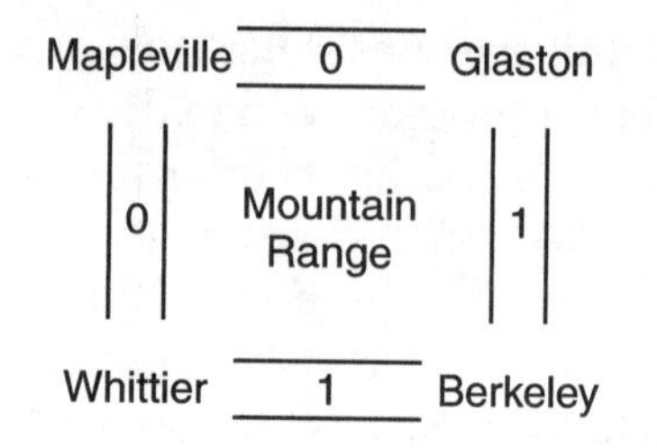

Now we can describe the road situation using numbers:

From	*To*	*Road Situation*	*Number Description*
Mapleville	Glaston	closed	0
Glaston	Berkeley	open	1
Berkeley	Whittier	open	1
Whittier	Mapleville	closed	0

The use of zeros and ones to describe a situation is a way of *digitizing* a situation. What was described with English words, such as open and closed, has now been described by the numbers *zero* and *one*. These two digits are called *binary digits*—the word *binary* implying there are only *two* of them: specifically zero and one. The term *bits* is formed from the Italicized letters of the two words "*bi*nary digi*ts*."

Now let's expand on the situation of the roads. What if someone from a nearby town inquires, "Can I get from Mapleville to Glaston and from Glaston to Berkeley?" (The towns surround a mountain range that prevents driving directly from one town to the other.) The answer would be "No, but you could go from Glaston to Berkeley."

MAPLEVILLE	→	GLASTON	→	BERKELEY
	closed		open	

If we digitized the answer, it would look like this:

MAPLEVILLE	→	GLASTON	→	BERKELEY
	0		1	

If we dropped all English words, the answer would be the two bits:

0 1

For the next question someone asks, "Can I go from Whittier to Glaston via Berkeley?" The answer would be yes, because the road from Whittier to Berkeley is open as is the road from Berkeley to Glaston."

WHITTIER → BERKELEY → GLASTON

open open

Now we digitize our answer:

1 1

The first one represents that the road is open from Whittier to Berkeley and the second one represents the road being open from Berkeley to Glaston.

Now look at an expanded map of the area—this time adding the names of smaller towns between the larger towns.

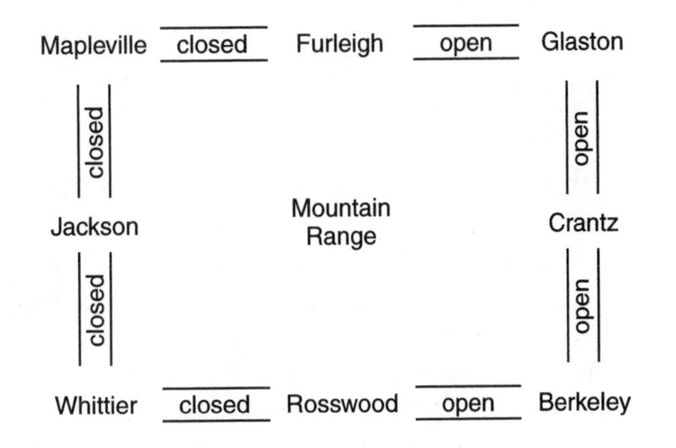

From	*To*	*Road Situation*	*Digit Description*
Mapleville	Furleigh	closed	0
Furleigh	Glaston	open	1
Glaston	Crantz	open	1
Crantz	Berkeley	open	1
Berkeley	Rosswood	open	1
Rosswood	Whittier	closed	0
Whittier	Jackson	closed	0
Jackson	Mapleville	closed	0

A description of the road situation from Mapleville to Berkeley via Glaston would look like the following:

MAPLEVILLE →	FURLEIGH →	GLASTON →	CRANTZ →	BERKELEY
	closed	open	open	open
	0	1	1	1

or just the four bits:

0 1 1 1

If we wanted to describe the road situations starting at Mapleville and going through Furleigh, Glaston, Crantz, Berkeley, Rosswood, Whittier, Jackson, and back to Mapleville, we would be looking at these bits:

0	1	1	1	1	0	0	0
bit	bit	bit	bit	bit	bit	bit	bit

one byte

Any sequence of 8 bits is called a *byte.* So the sequence we just used is a byte of information.

Binary digits or bits can be used to describe many situations in the real world aside from the road conditions mentioned. When these numbers or digits are used to represent a particular situation or bit of information, we have digitized information. What other examples of digitization are there? We digitize photos when we scan them—all the colors used to create images that we recognize are represented digitally. The music you hear on a CD has been digitized because musical notes can be associated with numbers as well. A password for a person using a bankcard at an ATM can be digitized. In Chapter 2 we will discuss how letters and words are digitized.

HINT!

Digitizing information means using numbers to represent something other than a number. We digitize sound, color, and even images. Once digitized, materials can be handled by a machine that can recognize these two integers: 1 and 0.

EXERCISES

1. Using the numbers zero and one, describe the road situation for a trip from Berkeley to Mapleville through Whittier.

2. If the weather clears up and all the roads are opened, how would you describe a trip starting and ending in Mapleville and going through all the towns surrounding the mountain range?
3. Give an example of something you use every day that involves the digitization of information.

1.4 THE COMPUTER: AN ELECTRONIC MACHINE

In order to be a good programmer, it is important to start with some of the fundamentals of how the computer operates. You must understand what kind of machine you're using and why it can do what it does.

The computer is an electronic machine. Thus it needs electricity to operate. The electricity comes from batteries or a plug in a wall and travels through elaborate circuitry etched onto highly sensitive material (like silicon) on a chip.

ELECTRICITY: ON OR OFF

By manipulating electricity through the circuitry on a chip, we can go back and forth between two states: electricity flowing and electricity stopped.

If we assign the zero bit (0) to the stopped state and the one bit (1) to the flowing state, these bits can be connected to the internal workings of the machine.

The machine's capacity for manipulating electronic states is used to create a language of zeros and ones called *machine language*. It is called machine language because it is the elementary language of the machine. Without getting bogged down in the details, the computer expresses all information with these bits and bytes of machine language.

1.5 COMPUTER LANGUAGES

Languages are used in everyday life for communication. All of us speak our native language and, depending on how much we read or write, we can develop a deep understanding of that language. The important thing about language is that you use it effectively to communicate your intentions, needs, wishes, and feelings.

Computer languages are similar to spoken languages in that you must use them very precisely so that you are not misunderstood. Each language has its own grammar or *syntax* which must be followed for the computer to understand that language.

Consider these examples from spoken languages:

English: Hello, how are you?
French: Bonjour! Ça va bien?
German: Guten Tag. Wie geht's?
Japanese: Konnichi wa. O genki desu ka?

All these examples mean the same thing: Each sentence has a greeting followed by a question asking how you are. But each example is a completely different group of words. Unless you know these languages, you would not know that each means the same thing.

Computer languages are similar in that there are basic tasks which any computer language must do for a programmer. The programmer just has to learn to "speak" the language. One advantage of a computer language over a spoken language is that is does not take that long to become fluent in a computer language! Many programmers learn several languages during their careers.

Now read these examples from some computer languages:

```
B.A.S.I.C.   if ( x > 5) print "greater"
Pascal       if x > 5 then writeln ('greater')
C++          if ( x > 5) cout << "greater"
Java         if (x > 5) System.out.println ("greater.")
```

All these statements accomplish the same task: If the contents in the variable called "x" is greater than the number "5," then we will print a message on the computer screen—the word "greater."

LEVELS OF LANGUAGE: HIGH AND LOW

All programming languages need to be translated into machine language, the native language of the computer. Machine language is comprised of binary code and, therefore, is tedious to read and understand. Because the computer can only execute commands that have been written in its native machine language, other languages must be translated into machine language before being understood by the computer. There are two levels of language among programming languages: *high level languages* and *low level languages*.

HIGH LEVEL LANGUAGES

High level languages are languages that are high above the language of the computer—its machine language—the language it "understands" most naturally. In order to understand the phrase *high level*, first consider these analogies:

Let's say you want to have a party for over 100 people. A "high level" party giver would simply pick a date for the party, invite friends via e-mail, or the telephone, and then the party would take place. A "low level" party giver would have to handle all the details (i.e., renting a place, hiring a caterer and a band, etc.). The party would not just "happen." The low level party giver would have to deal with all the details that the well-heeled "high level" party giver can ignore.

Suppose you need to get your car fixed. A "high level" approach to repair would be to bring the car to the repair shop and pick it up when the work is done. A "low level" approach would be to lift the hood yourself and examine the parts, looking for the problem. Once the problem part has been found, you would have to repair it yourself and then close the hood.

These analogies apply to programmers as well. The high level language programmer doesn't need to know anything about how the computer itself goes about its work. He states a problem at a higher level without being involved in the nitty gritty of how the computer performs its tasks. For this reason, high level languages use commands that relate directly to the problem being programmed as opposed to the internal workings of the machine.

Programs written in high level languages run more slowly on the computer because these languages need to be translated into machine language. Programs written in a high level language require fewer lines of code than those written in a low level language. It's much easier to say "let's give a party" or "fix the car" than to say "call the caterer, order the food, rent the hall, and the like." Pascal, Cobol, Fortran, B.A.S.I.C., C, C++, Perl, and Java are some examples of high level languages.

HINT!

Languages are defined in terms of their proximity to machine language: The higher the level, the more translation required before the program can be executed by the machine. Machine language is generated from the electronic states determined by the current and circuitry of the machine.

HINT!

Computers "understand" machine language: High level languages must be translated to that machine level so that the computer can "understand" them and execute their commands.

LOW LEVEL LANGUAGES

Low level languages are just above machine language level. As such, they needn't undergo as much translation as the high level languages. They are, however, more difficult to understand because they rely on a greater understanding of the internal workings of the machine. Assembly languages are low level languages.

LANGUAGE HELPERS: TRANSLATORS

Translators break down high level and low level language code into machine language understood by the particular processor in the C.P.U. There are two kinds of translators: *interpreters* and *compilers*.

INTERPRETERS AND COMPILERS

Translators can work in two different ways: as interpreters or compilers. *Interpreters* will translate one line of code at a time and generate error messages immediately. *Compilers* translate an entire file of code all at once, rather than line by line. The compiler will not generate error messages until all code has been translated. The original file or program that the programmer writes is called *source code. Object code* is the result of translation and is the machine language version of the original file. C++ is an example of a compiled language whereas B.A.S.I.C. is an interpreted one.

HINT!

Translators change high level language or low level language into machine language, which is readily understood by the computer's processor.

1.6 THE ALGORITHM: THE BASIS FOR ALL DESIGNS TO SOLUTIONS OF PROGRAMMING PROBLEMS

We are almost ready to begin the subject of programming—taking problems in the real world and writing them in a language that the computer can "understand" or execute. Before you get to the stage of programming your problems, you must first design a suitable way of solving problems. An *algorithm* is a set of steps for solving a problem. These steps may repeat and may involve some decisions such as a choice of two or more things.

Consider the following example of an algorithm for buying a ticket to a movie:

1. Go to the theatre.
2. Walk to the ticket counter.
3. Select a movie.
4. Pay the price.
5. Receive the ticket.

What about an algorithm for finding the smallest number among three numbers?

1. Compare the first number with the second number.
2. Discard the bigger number from step 1.
3. Compare the third number with the number that's left.
4. Discard the bigger number from step 3.
5. Whatever number is left is the smallest of all three.

THE THREE PARTS OF ANY ALGORITHM

The algorithm has three parts:

1. The steps are finite (i.e., they do not go on forever).
2. Steps may be repeated.
3. Steps may involve decision making.

HINT!

Each step of an algorithm should follow the step before. If necessary, repeat some of the steps and skip others if a decision calls for that action.

EXAMPLES OF ALGORITHMS IN THE REAL WORLD

Some examples of other algorithms follow.

Buying a bag of fries at Burger King:

1. Select the size of fries that you want.
2. Take out enough money to pay for them.
3. Hand over the money.
4. Receive change if necessary.
5. Take fries with you.

Dialing a friend who lives locally:

1. Pick up receiver.
2. Dial 7 digits.
3. Wait for friend's voice.

Programming a VCR's timer to tape a show:

1. Select menu button on remote control.
2. Select program menu from the given menu.
3. Complete table information about program's start time, finish time, channel, and speed of programming.
4. Press menu button on remote control.
5. Press power button of VCR.
6. Check for red light on VCR.

EXERCISES

What are the steps needed for an algorithm to do each of the following?

1. Put three numbers in order from least to greatest.
2. Set an alarm clock.
3. Log on to an email server.

1.7 PROGRAMMING

Now we are ready to discuss the subject of *programming*—taking a problem or task and designing an algorithm to handle this task; then using a programming language to express that algorithm so the computer will be able to execute that code.

Most people think of programming as being just about code—the lines of symbols and words that all of us have seen if we have ever opened a book on programming. It is more than code, however: However, it is a way of thinking about a problem and designing a solution that can then be written in a programming language.

Look at this short C++ program that prints the message, "Could you please say that again?" 250 times on the machine:

```
# include <iostream.h>
# include <string.h>

int main ()
{ int x;
  string first_phrase;
  first_phrase = "Could you please say that again?";

     for ( x = 0; x < 250; x++)
        {
          cout << first_phrase << endl;
        }
    return 0;
 }
```

Here is another version of the same program in Pascal:

```
program printmessage (input, output);
     var
        x : integer;  first_phrase :string;
      begin
first_phrase := 'Could you please say that again?';

      for x := 0 to 250  do
       begin
       writeln (first_phrase);
       x := x + 1;
       end
     end.
```

Each of these examples could have been written in any language. The main point of these programs is that they both contain a *loop*, which is another way of saying that a particular task is being repeatedly executed.

The point of showing both of these programs is to focus on how they are similar rather than on their language differences. As you move through the chapters that follow, keep in mind that languages will always change to suit the development of technology. Certain concepts remain the same regardless of language. If you learn these concepts well, you will have no problem becoming a serious programmer.

SUMMARY

In this chapter we looked at the differences between hardware and software and we defined the principal parts of a computer. We then looked at an example of the digitization of information. Because the computer is an electronic machine, binary digits can be used to describe the electronic states of the machine. These bits are used in machine language. All other languages need to be translated into machine language so the computer can execute their commands. The algorithm is necessary in the design of solutions to problems that programmers will code in any number of languages.

In Chapter 2 we will look at the first major concept of programming—understanding variables.

ANSWERS TO ODD-NUMBERED EXERCISES

1.3

1. 1 0 0 0
3. Music CD, DVD, digital cell phone, digital camera

1.6

1. Take the first two numbers and find out which is smaller. Put it in slot #1.

Take the third number and compare it with the number in slot #1. Whichever is smaller should be put in slot #1.

2. Now compare the remaining two numbers. Whichever is smaller should be put in slot #2and the remaining number should be put in slot #3.

3. Open the application
   Type your name
   Hit tab
   Type your password
   Hit enter

CHAPTER

2

Variables: The Holders of Information

IN THIS CHAPTER

We cover the topic of variables as necessary holders for data. We examine variables by their types—that is, what kind of data they are intended to hold. We define programming statements, the necessary building blocks of programs. Next we cover how to introduce a variable into a program: variable declaration. Finally the topic of loading variables with data will be covered—how the variable is assigned its data.

- Variables as Holders of Data
- Types of Variables: Integer, Real, Character, and String
- Variable Declaration
- Programming Statements
- Assignment: By the Programmer or by the User

2.1 A PLACE TO PUT DATA

One of the computer's most treasured assets is its capacity to store and manipulate information. *Data* (a plural word) is another term for this *information*. Most programs that programmers write will need to put this data or information into some type of holder to manipulate it in some manner. We see the need to manipulate data everywhere in our society. Changes of address, phone numbers, new passwords for accounts, and editing manuscripts illustrate the need to manipulate information. Programs are constantly updated.

A programmer might want to edit some information, change it, print it, or perhaps send the data over the Internet. Data has to be put somewhere and programmers use holders for it.

Three examples of data:

- The number 365 to represent the days in the year
- The number –20 F to represent the Fahrenheit temperature in Alaska on a winter day
- The name of a favorite actor, "Tommy Lee Jones"

These three pieces of data, 365, –20 F, and "Tommy Lee Jones" are all pieces of information that a programmer would have to store in holders. When we use holders, we try to give them descriptive names so that readers of the programs can understand what we are trying to do in our program.

Let's devise some appropriately named holders for our data. We could put the number 365 into a holder called *days*, the number –20 F into a holder called *temperature*, and the name "Tommy Lee Jones" into a holder called *actor*. Now instead of referring to 365, –20 F, or "Tommy Lee Jones" directly, we refer to the holders—*days*, *temperature*, or *actor*—that a programmer uses to manipulate data that often changes or *varies* over time.

Suppose you want to change 365 to 366 because you are dealing with a leap year. A programmer would manipulate the holder *days* rather than the number directly. Should you want to change the name of your favorite actor to "Tom Hanks," you need to put the new name inside the holder *actor*. Likewise, to reflect temperature changes in Alaska, we use the holder *temperature* to alter the degrees.

One can never be certain whether data in a program will change right away or later on; one also doesn't know how often a piece of data will change during the running of a program. Having a holder that contains data facilitates the management of that data.

Some examples:

A programmer would write instructions to do the following:

1. Increase the number in *days* by one.

2. Get another name of an actor and put that name into the holder *actor*.
3. Change the value in the *temperature* holder to the current reading.

These examples illustrate the need for a programmer to manipulate the values through their holders, because the instructions can be written for the holders rather than the numbers themselves. The programmer thus controls the data via the holder name (Figure 2.1).

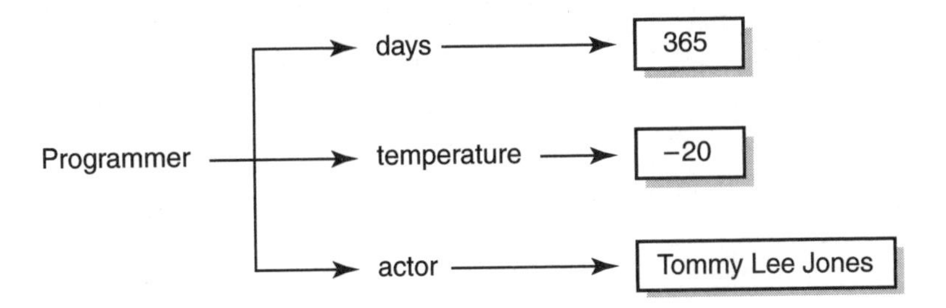

FIGURE 2.1 The programmer accesses data through the holder names used in his program.

Now to the real name of these holders—variables. *Variables* are holders of data, in particular, data we expect to change or *vary* over time. In computer languages, the word "variable" differs slightly from its meaning in algebra. Variable names are not simply the letters of the alphabet, such as 'x' and 'y,' which many of us remember from our study of algebra. Variables can be any *descriptive* names that help us understand *what the data means*. That's why we used variables named *days, temperature,* and *actor* to label or identify the raw data, 365, –20, and "Tommy Lee Jones."

WATCHOUT!

Variables in algebra are letters of the alphabet, such as 'x' and 'y,' but in computer languages, they can *also be* any *descriptive* names such as *sum, answer,* and *first_value.*

HINT!

If you want to link more than one word together to make an interesting variable name, use the *underscore* character '_'.

Some examples:

first_sum
My_last_answer

2.2 SOME EXAMPLES USING VARIABLES

This example shows why it is easier to write a program using a variable rather than a number/value. The *temperature* variable from the previous section could be used as a holder for a given daily temperature in Anchorage, Alaska during the last week of February. As the temperature changes each day, a new value is put inside of the *temperature* holder or variable.

	Sun.	Mon.	Tues.	Wed.	Thurs.	Fri.	Sat.
Temperature	15	18	6	21	19	−4	2

One way to view what a variable might look like as its values are changing is to show the variable and its values as they are being replaced. Each new value replaces the previous one until the final value is left there (at least for now!).

Temperature
15
18
6
21
19
−4
2

If a programmer wanted to write a program to print the average daily temperatures during the first four days of the week he could devise two algorithms to do this—one using the variable *temperature* and one without that variable.

Algorithm with *Temperature* as the Variable	Algorithm without Variable
1. Enter a value in *temperature*.	Print 15.
2. Print actual *temperature* for second day.	Print 18.
3. Print actual *temperature* for third day.	Print 6.
4. Print actual *temperature* for fourth day.	Print 21.

The algorithm *without* the variable is inefficient and dependent on exact values being known at all times. The algorithm using the variable *temperature* is *not* dependent on knowing what value is within *temperature*. It is enough to know that whatever the temperature is, it will be printed.

AN EXAMPLE OF THE COMPARISON OF TWO VARIABLES

In another example, we will use two variables and continually compare one value against a second value in terms of whether the first value is less than, greater than, or equal to the second value. Think of a flight of stairs with one variable representing the number of steps you have climbed and the other variable representing the top step. The first variable will be called *count_steps* and the second variable will be called *top_step* (Figure 2.2).

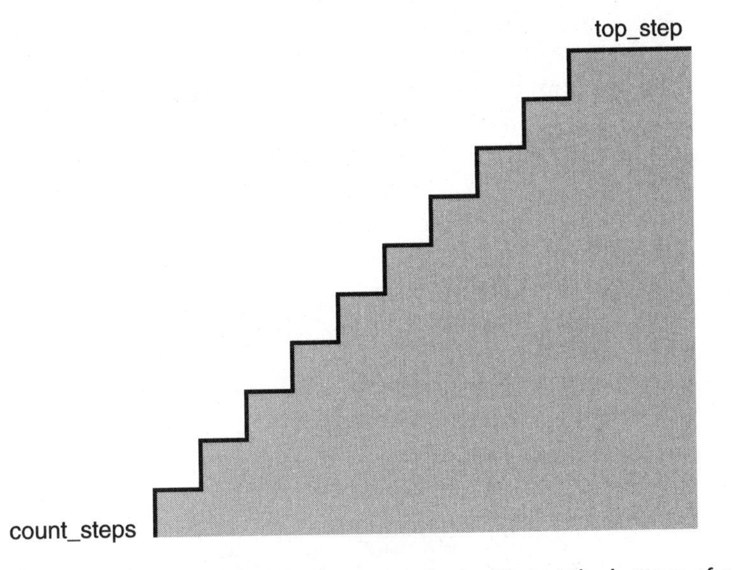

FIGURE 2.2 *Count_steps* is shown on the landing at the bottom of a flight of stairs and *top_step* is shown at the top of the steps.

We will continually increase the value of *count_steps* and each time check its value against *top_step's* value. If *count_steps* is still less than *top_step*, it means we have not yet reached the top of the stairs, and so we should continue to increase *count_steps* by 1. When *count_steps* and *top_step* are equal, we know that *count_steps* has "reached" the top of the flight of stairs and the program should end.

In this example, the number 10 is placed in the variable *top_step* . The value 0 is placed into *count_steps* to represent the fact that we are at the bottom of the flight of stairs.

Look at the values of *count_steps* at the beginning of the program.

count_steps

0	This statement reports that *count_steps* starts off as zero.
1	*Count_steps* changes to 1.

Now *count_steps* is compared with *top_step* in terms of being less than, greater than, or equal to *top_step 's* value.

count_steps	<	*top_step*
1	<	10
	"less than"	

Because *count_steps'* value is less than *top_step's* value, the execution of the program continues; *count_steps* increases to two.

count_steps	<	*top_step*
2	<	10
	"less than"	

This pattern continues:

count_steps	<	*top_step*
3	<	10
	"less than"	

count_steps	<	*top_step*
4	<	10
	"less than"	

This pattern continues until we view the program near the end of execution.

count_steps	<	*top_step*
9	<	10
	"less than"	

Until now, the variable *count_steps* has been less than the *top_step* of 10. But now:

count_steps = *top_step*

10 = 10

"equals"

The last value for *count_steps* is 10 and it is compared with the value in *top_step* .

Because *count_steps'* value is not less than *top_step* but, rather, equal to it, *the program stops.*

This example illustrates the ability of the computer to alter one variable repeatedly. The decision to alter this variable depends on its value in comparison to another fixed value, *top_step*. As the program executes, the focus shifts back and forth between *count_steps* and *top_step* with *count_steps* steadily increasing as the program runs. In one programming statement, *count_steps* is increased and in the next statement it is compared to *top_step*. The programmer uses this algorithm to do all the work.

Here is an algorithm to count the number of steps and then ring a bell when you get to the top:

1. Set *count_steps* to zero.
2. Set *top_step* to 10.
3. Increase *count_steps'* own value by 1.
4. Check *count_steps'* value against *top_step's.*
5. If *count_steps* is less than *top_step* go back to step 3; otherwise, go to step 6.
6. Ring a bell.

ON THE CD-ROM

A program that increases *count_steps* and compares it with *top_step* is written in C++.

2.3 DIFFERENT TYPES OF VARIABLES

Now that we have both introduced the term *variables* and looked at some examples involving them, it is important to classify variables according to their *type*. The *type* of a variable is the "kind" of holder needed for the kind of data being stored *inside* the variable. In our previous examples from the previous section, *days* and *temperature* were holders for the same kind of data: a *number*. The *actor* variable is a different type because it contained three words ("Tommy Lee Jones"). So what types of variables are there in most computer languages?

The main division in data types is between numbers and letters. Data is then further subdivided into groups under those headings (Figure 2.3).

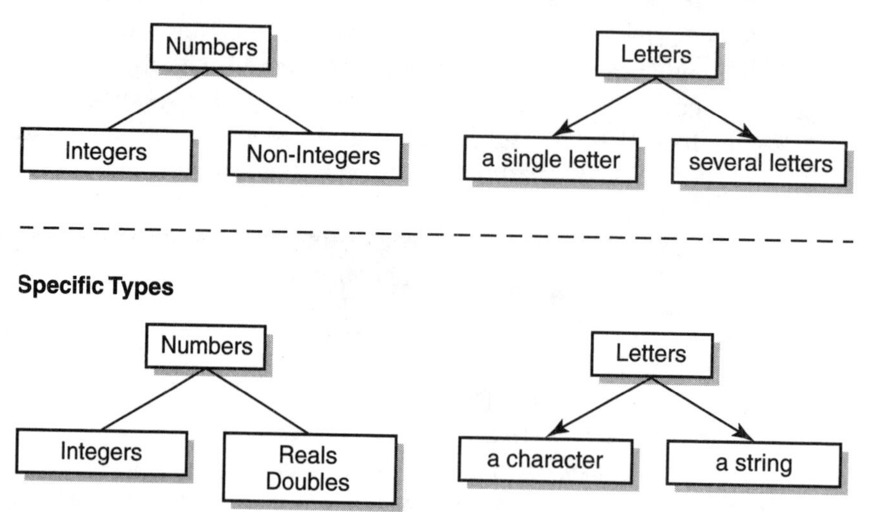

FIGURE 2.3 In the first diagram, the four main groups of data types are listed under their appropriate headings. Then another diagram gives the specific names of these four main groups.

THE INTEGER TYPE

Integers, in computer languages, are defined as numbers that have no fractional parts.

Some examples of integers:

–20
42
13
1,475
–234
0

THE REAL TYPE

Numbers that are not integers can be called *real*. In the computer language C++, real numbers are stored in a type called *double,* which refers to the fact that *two* bytes of memory are needed for its storage.

Each of the numbers in the following list has a fractional or leftover part, so to speak. This puts them in our second category, the real type:

14.62
58 ⅓
– 5.76
0.213
17.36
8.0

The term *real* refers to all other numbers that are not integers:

14.62
58 ⅓
–5.76
0.213
17.36
8.0

Even the 8.0 would be considered a real number because it was typed with a .0 added to it. The computer, therefore "thinks" it is a real number and not an integer. Just like a calculator, the computer will display these numbers as decimals on your computer screen:

14.62
58.333
–5.76
0.213
17.36
8.0

THE CHARACTER TYPE

Now consider the *character* type. The *character* type is a variable that will hold a letter of the alphabet or a symbol found on your keyboard like '#', '*', '!', and the like. Which symbols or letters are considered characters? There is a standard list of characters in the American Standard Code for Information Interchange, also known as the "ASCII" (pronounced "askey") chart. It includes the alphabet in both upper and lower cases as well as all the symbols on a keyboard. If you use any of these letters or symbols, you will need a character type variable to hold it. For a copy of the ASCII chart, please see the Appendix. Examples of non-ASCII (or extended ASCII) characters would be special letters from foreign languages, 'ç,' for example, or special mathematics symbols such as 'π.'

Examples:

'G'
'%'
'+'
'k'

HINT!

Characters are generally displayed with single quotations to avoid confusion with variables that have the same name as a character. In the previous example, the letter G could be the variable named "G" or it could be the letter 'G.' By putting single quotes around it, we emphasize that we are talking about the letter 'G.'"

THE STRING TYPE

The *string type* is a variable holder that can only contain *strings* or *groups* of letters or symbols. The string type allows words to be stored by the computer. The ability of a user to enter a word is via the string type. Strings are used for a sequence of characters because the character type can only hold one character at a time. Later we will see that there are other ways to handle a sequence of letters.

Examples:

"Washington, D.C."
"hello!"
"My name is Jack."
"^&*%($*."

Think of a string as a string of beads in a necklace. Except here we mean a string of characters strung together (Figure 2.4).

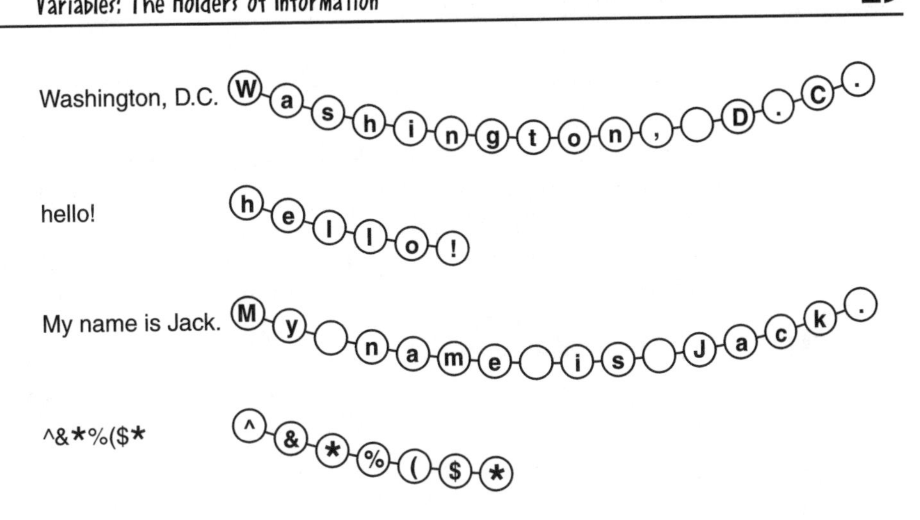

FIGURE 2.4 Each example is shown as a string of beads in a necklace to emphasize that each letter is separate but also connected to the next letter, as far as the computer is concerned.

HINT!

The word "type" in computer programming languages refers to the kind of holder needed for data.

TYPE	EXAMPLES
Integers	1, 2 ,–38, 327, 10, etc.
Reals	3.82, –14.6, 0.005, π
Characters	'G', '%', '+', 'k'
String	"robot," "Stars and Stripes,"" "Wow!!"

2.4 INTRODUCING A VARIABLE FOR THE FIRST TIME IN A PROGRAM

Now for an explanation of the most interesting and basic part of programming—the ability to store data/information in a variable. The initial step in writing a computer program is to tell the computer how much memory to *allocate* or *set aside* for the variables used in the program.

The computer must know whether the variable being used requires a lot of *memory* or space, for instance. Once the computer knows what type of variable is being used, all the rules that govern that particular type must be followed. These rules will be explored later as we learn more about programming.

To tell the computer what type of variable will be used and how much memory is required, we must declare a variable. *Declaring* a variable is the same as *introducing* a variable. It is the first time you tell the computer what the variable's part will be in the program. The relationship or type of a variable must be identified so the computer can respond appropriately. The different types of variables we have mentioned—integer, real, character, string—all make different demands on the computer. For example, different types require more memory than others. A character requires less space for storage than does a string. Likewise, an integer requires less memory than does a real type on most computers. By stating the type of variable used in the program, a programmer is instructing the computer to allocate a certain amount of memory for that variable.

AN ANALOGY: TELLING THE AUDIENCE WHO IS WHOM

The analogy of a playbill is useful because we are likening the computer to an audience. An audience wants to know who is playing what part as well as the kinds of roles in the play. The computer is the same way: It needs to know what part each variable is playing so it can respond accordingly. Variables have different memory requirements for storage, and the manner in which they are stored depends on their type. A computer has to "know" this before the variable appears later in the program.

Think of the play, "Hamlet," by William Shakespeare. When one opens the playbill to the first page, the list of characters and their parts in relation to the whole are listed. Look at the first page of "Hamlet" (Figure 2.5).

Hamlet is a prince, whereas Rosencrantz and Guildenstern are courtiers. Lesser known Francisco is a soldier. Gertrude is a queen and Hamlet's father is a ghost. A playbill lets the audience know who's involved in the play and what part each character plays by the description of the character. From reading the cast of "Hamlet," we know that Gertrude is a member of royalty, whereas Rosencrantz and Guildenstern are not.

Cast of Hamlet

Claudius, the new king of Denmark.
Hamlet, son to the late, and nephew to the present king.
Fortinbras, prince of Norway.
Polonius, a lord and high official (probably Lord Chamberlain).
Laertes, his son.
Horatio, the friend of Hamlet.
Voltimand
Cornelius
Rosencrantz } Courtiers
Guildenstern
Osric
Marcellus, a Danish officer.
Francisco
Bernardo } soldiers on sentry duty.
Reynaldo, servant to Polonius.
A Norwegian Captain.
Players on Tour.
Two clowns, gravediggers.
English ambassador, a priest, a gentleman, soldiers, sailor, messenger, and various attendants.
Gertrude, Queen of Denmark and mother of Hamlet.
Ophelia, daughter of Polonius.
Ghost of Hamlet's father.

FIGURE 2.5 This page shows the entire cast of "Hamlet."

Imagine a playbill with this heading:

"My First Computer Program!!"

Starring:

The integer	*my_first_sum*
The integer	*my_last_sum*
The real	*answer*
The character	*middle_initial*
The string	*last_name*

The "type" of each variable is written on the left, while the "name" of the variable is in italic letters on the right. This playbill is useful because we can find out that there are two integers in this program—one called *my_first_sum* and another called *my_last_sum*. There is one real called *answer*, a character called *middle_initial*, and a string called *last_name*.

Each of these "players" has a proper name, a name that the audience would recognize. *my_first_sum*, *my_last_sum*, *answer*, *middle_initial*, and *last_name* are the names of the "players" in this program. They take part in the program and appear at different points in the program to perform different tasks. The fact that one of the variables is an integer, while another is a real is comparable to saying that one of the characters in the play "Hamlet" is a queen, while another character is a maid.

This "play," "My First Computer Program!!" also has two integers in it: *my_first_sum* and *my_last_sum*. These names are useful because they allow you to distinguish one integer from another. Most programs have several variables and often a few from each type category. Using the proper name to refer to the variable is a way of distinguishing among variables of the same type. This *proper name* is called the *identifier* of the variable to distinguish it from its *type*.

Look at this chart that mixes some of the types from both the program and the play. Each character in the play has a "type" just as each variable does. Is the character in the play a friend of Hamlet's, a relative, or a member of the ruling class? Is a variable an integer, a character, a string, or a real?

CHARACTERS IN PLAY		VARIABLES IN PROGRAM	
King	*Claudius*	Integer	*my_first_sum*
Queen	*Gertrude*	Integer	*my_last_sum*
Courtier	*Rosencrantz*	Real	*answer*

TYPE	IDENTIFIER
Queen	*Gertrude*
Prince	*Hamlet*
King	*Claudius*
Integer	*my_first_sum*
Integer	*my_last_sum*
Courtier	*Rosencrantz*
Courtier	*Guildenstern*
Real	*answer*

All programs require that variables be *declared*; that is, *their type and name are stated before* they are used in the program to do something really useful. Imagine what would happen if you were watching a play and an unidentified character appeared in Scene 3, for example, and spoke with other characters before leaving the stage. Your first instinct would be to check the playbill and make sure you had not overlooked this character. However, what would you think if the character wasn't listed there? You would have to spend the rest of the play trying to figure out what connection that part had to the other parts in the play and the play as a whole.

The computer is the same way. It cannot encounter a variable halfway through a program unless it has been declared prior to use. This is one of the most important concepts that must be understood. It's not that all languages require that all variables be listed at the top of a program (though many do!), but they must at least be declared prior to being used in the program.

HINT!

Names for variables, as essential as the names of characters in a play, should be *descriptive* to enable any reader to follow what the variable will be used for in the program (e.g., *check_sum*, *last_name*, *initial_balance*, etc.).

WATCHOUT!

Don't underestimate the importance of declaring a variable because the computer *must know* how much memory to set aside for it. This must be done *before* the variable is used in the program.

2.5 A WORD ABOUT STATEMENTS

Computer languages, like spoken languages, have a certain grammar that must be followed. The first point about writing in a programming language is to write in sentences or statements. Computer *statements* are the *building blocks* of programs just as sentences are the building blocks of paragraphs and essays. A program is comprised of statements. There are different kinds of programming statements: *loop* statements, *if... then* statements, *assignment* statements, and *print* statements—just to name a few.

A statement in a computer language has rigorous grammar. As we learn each of these statements, we will take a moment to learn the *grammar*, often called *syntax,* of each statement type. This will save us a lot of time when we start to run programs. There is little or no room for variation because the "reader" of our programs is a machine that will not understand what we mean to say if we do not expressly say it according to the grammar it "understands."

TERMINATION OF STATEMENTS

The first point about programming is to understand how your computer language ends its statements. Do the statements end in a period (.) or a semicolon (;)? Most languages use the *semicolon* to indicate the end of a statement.

Let's look at these statements from various languages. It is not important that you understand the statements—only that you recognize that each one ends in a *semicolon*.

```
Example 1.   c := 14 ;
Example 2.   answer = 58 ;
Example 3.   if (x < 14)
             cout << "hello!\n";
```

In each of these examples, the semicolon is the last mark of punctuation after the group of words and symbols in the statements. In Example 3, we have a statement that wraps to the next line, indicating that programming statements do not need to fit entirely on one line.

HINT!

Programming statements terminate with some sort of punctuation; usually it is the semicolon.

MUST STATEMENTS FIT ON ONE LINE?

A programming statement need not fit on one line. Sometimes one statement, because of its complexity, will last several lines and the semicolon will appear finally to indicate that the statement is complete.

Another point about programming statements is that programmers like to indent. Indentation makes the programs easier to read and understand. In Example 3, the line:

```
if ( x < 14 )
cout << "hello!\n";
```

could have fit on one line but writing it on two lines makes it easier to understand as you will see after studying this kind of statement.

TIP!

A programming statement is not defined by its length. It may be longer than one line or it may wrap to the next line. Lines that wrap to the next line are easier to read and understand.

2.6 PUTTING A VALUE INTO A VARIABLE

Once variables have been *declared*, we are ready to put a value into the variable: This is called *assigning the variable*. By assigning the variable, we are giving it a value for later use. If we try to print the contents of a variable on the screen, and that variable is unassigned (or empty), then we have a problem.

WATCHOUT!

Sometimes programmers forget to assign a variable and then later in a program they try to use the variable's value. Remember to assign the variable before using it.

There are different ways of assigning a variable depending on who does the assigning of values. So now we formally introduce two important people in programming: the *programmer* and the *user*.

THE PROGRAMMER

The programmer is the person who writes the program. When a programmer puts values into variables, he is the only person controlling what data is assigned to those variables. The programmer might use his own data to load the variables or he might use an external source for that data, such as a file. But the bottom line still holds; he is responsible for acquiring the data and assigning it to the variables.

THE USER

The *user* is the person who interacts with the program. He will use the program much like the person who purchases an application (like Microsoft *Word*™) at the store and uses it after installation on the computer. The user runs the program and responds to what the program asks. If the programmer has written statements to cause the program to pause and wait for a response from the user, then the *user* will be the one to assign variables. The user is *prompted* or *asked* by the programmer with some message that appears on the screen (Figure 2.6).

```
Please type a number
|
```

FIGURE 2.6 A message followed by a "blinking" cursor appears on the screen. Whatever the user types (in response) will go into the variable associated with this prompting statement.

HOW THE PROGRAMMER ASSIGNS VARIABLES

In an *assignment statement,* a programmer writes a variable followed by some symbol to denote that a value is being assigned (sent into) a variable. To the right of the symbol is the value itself or another variable.

The syntax is as follows:

VARIABLE	*ASSIGNMENT SYMBOL*	*VALUE ;*
left-hand side	=	right-hand side

Look at each of the following assignment statements written in their respective languages:

VARIABLE	*ASSIGNMENT SYMBOL*	*VALUE;*	*(LANGUAGE)*
A	:=	35 ;	*(Pascal)*
A	=	35 ;	*(C++)*
Let A	=	35	*(B.A.S.I.C.)*

Everything on the left-hand side is the variable "A" while everything on the right-hand side is the value 35. The value 35 is being placed into a variable called "A." The emphasis in assignment is that some value from the *right-hand side* is being *shifted into the left-hand side* of the assignment symbol. In the examples that follow, different values are assigned to different types of variables.

Another important rule about assignment is that you must be careful about what values you put into which types of variables. In general, you must put a value into a variable of the same type. (*Integers* go into *integer type* variables; *reals* go into *real type* variables; *etc.*)

Some examples:

answer = –14; sum = 27 ;	integers are being assigned to integer variables
first_initial = 'M' ; last_initial = 'W';	characters are being assigned to character variables
name = "Janet"; friend = "Mike";	strings are being assigned to string variables
balance = 1234.56; amount = 78.00;	reals are being assigned to real variables

Before the assignment statement is executed, a variable is "empty" and has no contents. Once the assignment statement is executed, the value is put into the variable (Figure 2.7).

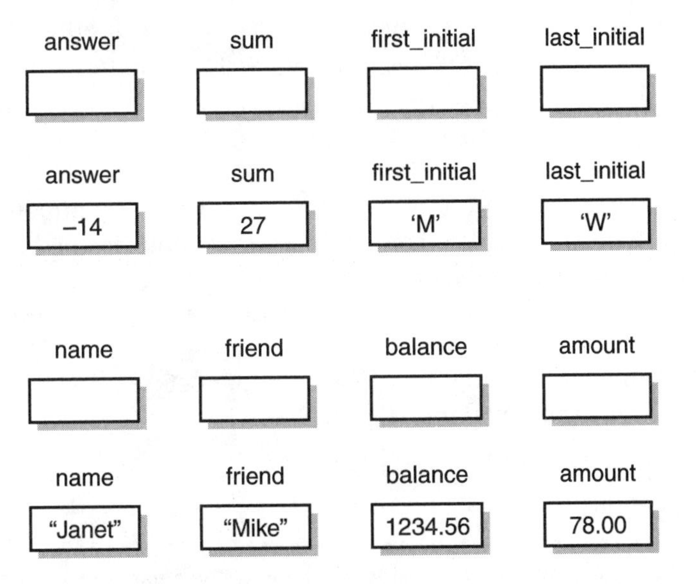

FIGURE 2.7 Each variable is shown initially empty or "unassigned" and then is shown with its value inside it.

WATCHOUT!

The one exception is that *integers can fit into real types* because integers are easily converted to real numbers and the memory required for storing a real is generally larger than that required for storing an integer.

EXERCISES

Identify the type holder on the left-hand side given the values being assigned from the right-hand side of the assignment symbol.

1. c := 67;
2. my_name = " Elsa";
3. marker = '$';
4. second := -128.45;

ASSIGNMENT OF ONE VARIABLE TO ANOTHER VARIABLE

It is also possible to *assign to a variable* the contents of *another variable*. Let's say you want an extra copy of a value which already exists in a variable, called *first_val*. All you have to do is declare another variable called *copy_val* and assign to *copy_val*, *first_val's* value. The syntax is as follows:

VARIABLE	*ASSIGNMENT SYMBOL*	*VARIABLE ;*
left-hand side	=	right-hand side

The variable on the right-hand side of the assignment symbol will have the value *inside it* copied into the variable on the left-hand side of the assignment symbol. If the variable is not empty, whatever is in it will be *replaced* with the new value. *This is called assigning one variable to another* (Figure 2.8).

In another example, a variable is empty before being assigned the contents of a second variable (Figure 2.9).

WATCHOUT!

The left-hand side must always be a variable, whereas the right-hand side can either be a value or another variable.

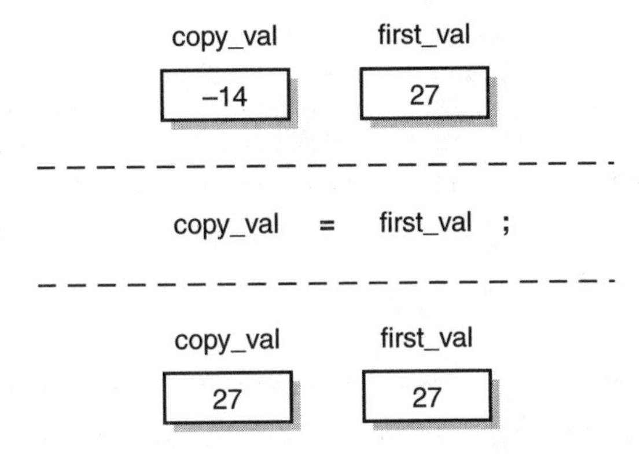

FIGURE 2.8 Variables *first_val* and *copy_val* are shown with their respective values. Then the assignment statement is executed. Now the variables are shown again—this time *copy_val's* value (–14) has *changed* to show the same value that is in *first_val* (27).

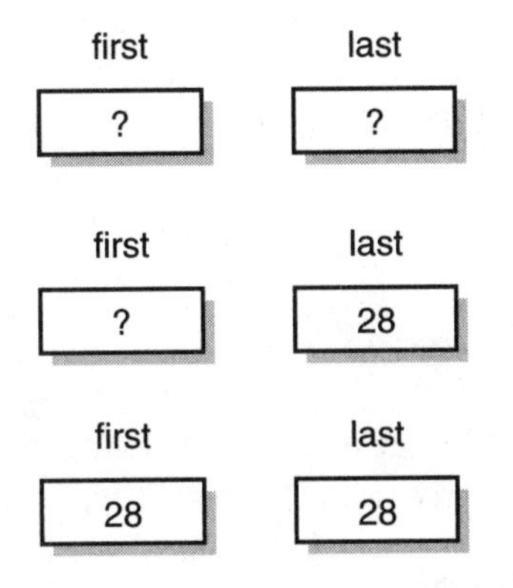

FIGURE 2.9 The variable *first*, initially empty, is assigned the contents of the variable, *last*. So, *first* is shown as an empty variable and *last* with its value, 28. Then the assignment statement is executed after which *first* has the same value as *last*—28.

HINT!

When one variable is assigned to another, the right-hand variable does not change. Only the contents of the left-hand variable change.

EXERCISES

1. Assign the value –28 to a variable called "C."
2. Assign the value 14.3 to a real called "D."
3. Assign the contents of "C" to a variable called "E" of type real.
4. What type of variable is "C"?

2.7 HOW THE USER ASSIGNS VARIABLES

It is important to let the user think of values for variables. The programmer should not be the only one thinking of values that will be assigned to variables. Think of the word "input" in English. Sometimes a friend will ask you, "What is your input?" He is really asking, "What do you think about (something)?" When we look for the user's *input*, we are interested in *what the user wants* for values.

There are two steps involved in understanding how we can get a response from the user stored into a variable. The first step is understanding how programming languages use the keyboard as a source of input.

INPUT STREAMS

Programming languages use an *input stream* as a source of values. Think of a stream in a computer language in the same way as you would a stream or rivulet (a small river). Streams flow into larger bodies of water. In programming languages, an input *stream* is like a small *channel* that *feeds from the keyboard*. Anything that is typed at the keyboard will be sent to this stream. (It is also visible on the screen.) The stream analogy continues as the stream is then a source of values (data) for the variables that will be assigned (Figure 2.10). Variables are assigned their values from this stream.

Programming languages will differ in the way that they handle the stream. One thing is common, however, and that is that every programming language uses some word to indicate that an input stream is the source of values for variables that will be assigned. The programmer must also *list* the variables that are to be assigned values from this stream. Let's concentrate on the C++ programming language, which uses its input stream name, "cin," directly.

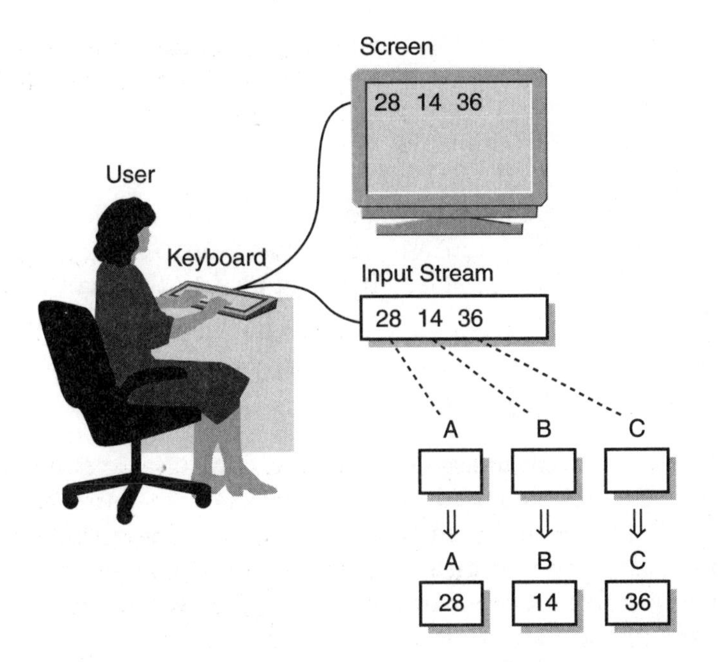

FIGURE 2.10 A drawing shows the connection among the user, the keyboard, the screen, the input stream, and the three variables being assigned.

A STREAM USED FOR INPUT: CIN

"*Cin*" (pronounced "see-in") is the name of the input *stream* in the C++ programming language. When you use a "cin" statement with a list of one or more variables, you are assigning data from the "*cin*" channel to the variables that follow. Anything that was sent to this stream will eventually end up in the variables connected to the stream.

The syntax for using the *cin* stream goes as follows. You name the stream, cin, then you follow with an *extraction operator* that indicates that something is being *pulled out* of the stream and sent into a variable. (The word *extraction* means "pull out"; that's why when you get a tooth pulled by the dentist, it is called an extraction!) The extraction operator is this symbol: >> . Next, you list the variable that you wish to load (assign) with the value that came from the stream. Let's look at some examples in the C++ programming language.

```
int second_val;
cin >>  second_val;
```

```
string my_name;
cin >>  my_name;
```

In the first example, we are taking a number (an integer) from the "cin" stream and assigning it to the variable, *second_val*. In the second example, we are taking a word (typed at the keyboard by the user) from the stream "cin" and assigning that word to the variable *my_name*. Notice how the variables always appear to the right of the extraction operator (>>).

ASSIGNING TWO VARIABLES AT ONCE

When you wish to assign more than one variable at a time with input from the user, you need to use the extraction operator *before* each variable in a list. We use the cin stream followed by the extraction operator, the first variable we are assigning, the extraction operator (again), and the second variable. (Remember we are assigning two variables with two numbers "extracted from" the stream.)

In both examples that follow, *more than one item* is pulled from the stream so the extraction operator is placed *in front of each* variable assigned.

```
cin  >>first_val  >> second_val;//two vars. are assigned

int a ; int b; int c;
cin  >> a >> b >> c ;//three vars. are assigned
```

WATCHOUT!

Any input from the user will first travel to the cin stream. You need the extraction operator, >>. If there is more than one number or word of input, put the extraction operator *before* each variable. You need a variable for each value and the variable should have the same type as the expected value.

HINT!

As the user types at the keyboard, she may make a mistake and need to backspace or delete. Her data will only become finalized after she presses the "return" key—then the data goes to the input stream.

EXERCISES

Identify who is assigning each of the variables—the user or the programmer.

1. m = 24;
2. cin >> first_value;
3. cin >> sum;
4. answer = -89;

SUMMARY

In this chapter, you learned that variables are the holders of data. Different *types* of variables were introduced: *integer, real, character, and string.* Next, the topic of declaration or how a variable is first introduced in a program was covered. If a variable is not introduced, the computer will not know what the variable's role in the program is. Furthermore, the computer will not know how much memory to set aside for that variable. Programs consist of statements that proceed in sequence like the previously mentioned algorithms. Statements are like sentences; both have punctuation to show that they have terminated. Most computer statements end in a semicolon. Each computer statement has its own grammar or syntax which must be followed rigorously.

Finally, the topic of assigning variables and who does that assignment, the programmer or the user, was discussed. If the programmer assigns the variables, we expect to see an assignment symbol surrounded by variables on one side and variables or values on the other side.

If the user assigns the variables, the user will type values at the keyboard. These values will be sent to an *input stream.* Variables are assigned values that come from this stream. This is how the user can assign values to variables.

ANSWERS TO ODD-NUMBERED EXERCISES

2.6 (i)

1. integer
3. character

2.6 (ii)

1. C = -28 ;
3. E = C ;

2.7

1. programmer
3. user

CHAPTER

3

Everything You Ever Wanted to Know about Operators

IN THIS CHAPTER

- Operators Defined
- P.E.M.D.A.S.
- Binary and Unary Operators
- Arithmetic Operators
- Relational Operators
- Logic Operators
- Mod Operator

Now that you have learned to store values into variables, what can be done with those variables? Most programs involve some form of computation: computing the balance in a checking account, comparing an identification number with an existing one to see if it is the same, or updating the information on an individual's age stored in a data file. To perform any kind of computation, we must first understand something about manipulating values within a computer language environment. *Chapter 2 told us how to put values into variables.* Now that we have those values inside of variables, we want to start *performing computations on those values. Operators* allow us to do that.

3.1 WHAT ARE OPERATORS ANYWAY?

Operators are the names for the *actions* that we perform on numbers. What usually comes to mind are the actions of addition, subtraction, multiplication, and division. Each of the following symbols: +, -, * (multiplication), and / (division) is an example of an *operator* because each acts on the numbers around it to generate an answer.

EXAMPLES

5 + 6 produces the answer 11
13 * 5 produces the answer 65
12/2 produces the answer 6

Not all operators are like the +, -, *, and / operators. Some operators are more subtle—they produce answers that are not numbers but *"true"* or *"false"* answers. The examples that follow use these operators: "<" (less than) or ">" (greater than).

15 < 37
"15 is less than 37"
Produces the answer: True

14 > 100
"14 is greater than 100"
Produces the answer: False

We get an answer, but it is not a number but, rather, a "true" or "false" answer.

The reason to use operators, namely + , - , * (multiplication), and / (division) is to enhance the ability of the machine to carry out simple tasks that involve some computation (e.g., computing the interest in a savings account, finding the average of a set of test scores, calculating the cost of an item that has been discounted, or checking to see whether someone's checking account balance is greater than the amount written on a check).

If you want to control this machine, it will be necessary to understand how computations, which involve *operators*, are performed. The easiest thing to do will be to understand the order of operations that already exists in the real world, namely those used in arithmetic and algebra.

3.2 THE ORDER OF OPERATIONS

Not all operators are the same. Some are more important than others in the sense that they are given priority over another operator. Here we use the term *precedence*. The *precedence* of one operator over another means that one operator *ranks higher* than another: It has *greater precedence*. If you use *more than one operator* in a programming statement, you will need to understand the precedence of your operators so that you get the desired result from any computations you wish the computer to perform.

The operator with the *highest rank/priority* will be executed *first* by the computer. Then any other operators in the statement will be executed in the appropriate ranking order. If we write an addition or subtraction operator and a multiplication or division operator in the same expression, the multiplication/division operator will be given *precedence*; that means that it will be executed *first* before the addition/subtraction operator.

Let's look at these examples where each operator has a number beneath it to indicate the order in which its operation is performed.

EXAMPLES

3 + 2 * 5

In this example, the multiplication is the ranking operation so 2 * 5 will be executed first—it is given precedence. Once performed (2 * 5 = 10), then the addition operator is activated (3 + 10) and the answer is 13.

Ranking Operators		1st		2nd	
	4	*	6	–	3
		⇓			
Becomes	24	–	3		
		⇓			
Which then becomes		21			

4 * 6 is computed first. Next, 3 is subtracted from 24 to equal 21.

Ranking Operators		1st		2nd	
	8	/	2	–	5
		⇓			
Becomes	4	–	5		
		⇓			
Which then becomes	– 1				

8/2 is computed first. Next, 5 is subtracted from 4 to equal –1.

P.E.M.D.A.S. "PLEASE EXCUSE MY DEAR AUNT SALLY"

Although the above phrase may be a little outdated, it continues to serve its purpose well— to give instruction regarding the precedence of the operators. The acronym, P.E.M.D.A.S., refers to the following: *P*arentheses, *E*xponents, *M*ultiplication, *D*ivision, *A*ddition, and *S*ubtraction. These operators should be executed in the order that they appear in the acronym. Parentheses are first, followed by exponents, then multiplication and division follow—with addition and subtraction last.

In the previous examples, the multiplication and division operators always preceded the addition and subtraction operators. Division and multiplication have equal precedence and they both rank higher than addition and subtraction. Likewise, addition and subtraction have equal precedence and they both rank lower than multiplication and division.

PARENTHESES STILL RULE

Parentheses are perhaps the most interesting operator because they tell the computer that what appears *inside of them* should be *given precedence* over all other operators (i.e., that they should be executed *first).*

EXAMPLES

Ranking Operators	8 (5 + 2) – 7
	2nd 1st 3rd
becomes	8 (7) – 7
	56 – 7
	49

Because the parentheses are present, (5 + 2) is executed first. Next, 8 is multiplied by 7 to equal 56. Then 7 is subtracted from 56 to equal 49.

15 + 3 (2 – 5) * 6
4th 2nd 1st 3rd
15 + 3 (–3) * 6
15 + –9 * 6
15 + –54
–39

WATCHOUT!

Remember that when the *same operator* appears *more than once* in a programming statement, each of those operators will be executed in order *from left to right*.

EXERCISES

What should be done first in each example?

1. 3 – 2(4 – 1)
   a) 3 – 2
   b) 2(4)
   c) 4 – 1
2. (5 + 7)4 – 8
   a) 5 + 7
   b) 7(4)
   c) 4 – 8
3. a(x + 7) – b (4 + 5)
   a) a(x)
   b) x + 7
   c) 7 – b
   d) b(4)
   e) 4 + 5

In each statement, write the precedence of each operator (the order in which it will be executed) underneath it. (The highest precedence gets 1, second highest gets 2, etc.)

4. 8 – 4 (5 + 1)
5. 3(2 – 5) + 6 * 6
6. 14 – 3 (10 + 2 * 4) – 1

What is the result from executing each of the following?

7. 8 – 4 (5 + 1)
8. 3(2 – 5) + 6 * 6
9. 14 – 3 (10 + 2 * 4) – 1

3.3 OPERATORS: ARE THEY BINARY OR UNARY?

Operators can be classified according to the *number* of *operands* required by the operator. An *operand* is a number on which an action is being performed. In the example, 2 + 2, the "two's" are the operands because the operation of addition is being performed on them. As you will see in the examples that follow, almost all operators require *two* operands. Any of our arithmetic operations, such as addition, subtraction, multiplication, and division are *binary* operations because you always need *two* numbers to add, subtract, multiply, or divide. (Recall that binary means two.)

BINARY OPERATORS

Binary operators are operators that require *two* operands, *one on either side* of the operator. The two operands are needed for the operator to function properly. Most of the operators you will encounter will be binary operators. Consider the following examples.

EXAMPLES

6 + 7
The '+' sign *needs two operands*, the 6 and 7, to work on.

3 * 8
The '*' sign *works on both the 3 and 8* on either side of it.

6 – 18
The '–' sign allows the 18 to be taken from the 6.

sum = 24;
The '=' sign allows the value 24 to be stored into "sum." It needs a variable on the lefthand side and either a number or a variable (containing a number) on the right-hand side.

UNARY OPERATORS

Unary operators are operators that require only *one operand* to function properly. The negative sign, '–,' can be an example of a unary operator. For example, if we put a "negative" sign in front of a number (5), it makes the number negative (–5).

EXAMPLES OF UNARY AND BINARY OPERATORS

1) *Unary*
– 4
The only operand is the 4. The result is the number, – 4.

2) *Binary*
3 + 2
There are two operands: "3" and "2."

3) *Unary and Binary*
– (8 + 7)

The '+' is a *binary* operator. Its operands are "8" and "7." The *unary* operator is the negative sign. Its operand is the 15 that results from executing 8 + 7 first.

HINT!

Remember the assignment symbol (=) from the last chapter? The assignment symbol is a *binary* operator because it always takes *two* operands: the variable to be assigned as the left operand and the value it is given as a right operand. (Some examples: a = 15; b = –34; and m = b).

HINT!

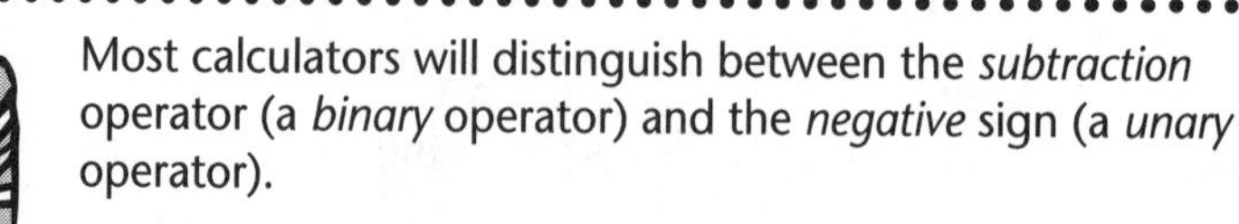

Most calculators will distinguish between the *subtraction* operator (a *binary* operator) and the *negative* sign (a *unary* operator).

EXERCISES

Underneath each operator, label it "binary" or "unary."

1. 38.7 + 56
2. –15 + 7
3. –(32 – 6)
4. answer = 78

3.4 ARITHMETIC OPERATORS

Every computer language will use the arithmetic operators to perform simple calculations. These operators are readily recognized as the same as those on a calculator, except, perhaps, for the multiplication symbol, '*'.

- \+ The addition sign
- – The negative sign or the subtraction sign, depending on how it is used
- * The multiplication sign
- / The division sign

DIVISION: A SPECIAL CASE

Except for the division sign, the other operators perform as they do in arithmetic. The division operator performs according to the type of the *operands*, those numbers used on either side of it (Figure 3.1).

Division works just as you would expect, *except when integers alone* are used with the division sign. If the division symbol is used with two integers, only an integer will be produced and any fractional part *will be dropped*.

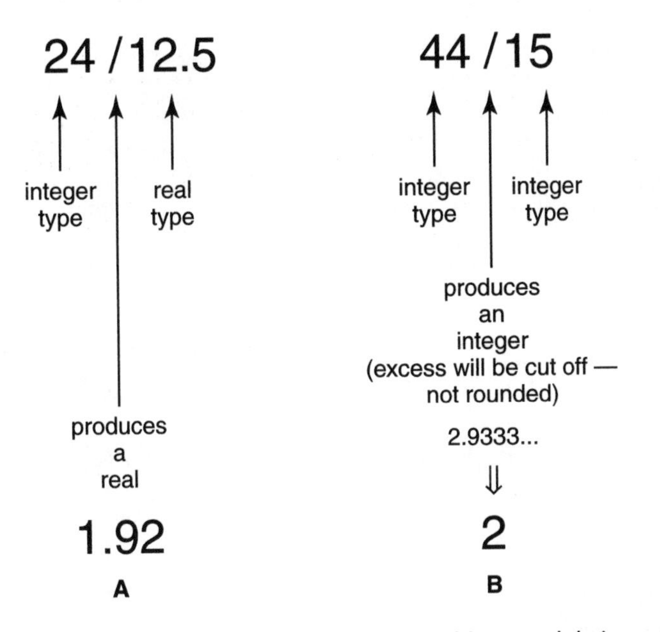

FIGURE 3.1 (A and B) Two division problems and their results are shown. If integers are used exclusively (3.1B), an integer will be produced.

EXAMPLES

6/4 produces 1 instead of 1.5
(0.5 is thrown away)

8/4 produces 2

–143/7 produces –20
(0.429 is thrown away)

EXAMPLES WITH VARIABLES

a = 14;
b = 12;
c = *a*/*b*;
(*c* is 1.)

answer = 67;
sum = –23;
new_answer = *answer*/*sum*;
(*new_answer* is –2.)

WATCHOUT!

Always notice which types you are using around a division symbol. The division symbol "recognizes" its *operands* and performs accordingly. Either the fractional part is *dropped* (with integers) or it is *retained* (with all others).

EXERCISES

What will be the value contained in each of the variables that are being assigned?

Assume each of these variables is of the *integer* type.

1. *answer* = 228/197
2. *sum* = 17/3
3. *m* = 135/5
4. *result* = –67/–3

Assume that each of these variables is of the *real* type.

5. *first_answer* = 325/12.6;
6. *last_answer* = –45/23.78;
7. *a* = 56.67/34

3.5 THE RELATIONAL OPERATORS

Another group of operators are very useful because they allow values to be *compared* to other values. When you compare one value to another, you are interested in knowing whether one value is *greater than, less than,* or *equal* to another value. The result of using a relational operator with its operands is a *true/false* answer *rather than* an *arithmetic* answer.

The first relational operators that we will use are as follows: <, <=, >, >=, = = . The last operator deserves some mention. The double equal symbol is used to emphasize that *this is not the assignment operator* from Chapter 2 (=), but the relational operator, *"equivalent to."*

EXAMPLES

	Result
15 < 16 15 "is less than" 16	*True*
–13 > 100 –13 "is greater than" 100	*False*
25 = = 25 25 "is equal to" 25	*True*
35 = = 4 35 "is equal to" 4	*False*

In addition to the < (less than), > (greater than) , and = (equal to) operators, there are the additional <= (less than or equal to) and >= (greater than or equal to) operators.

57 <= 69 57 "is less than or equal to" 69	*True*
12.5 >= 12.5 12.5 "is greater than or equal to" 12.5	*True*
–26 <= –27 –26 "is less than or equal to" –27	*False*

Finally, computer languages provide an operator that means "not equal to": ≠.

In the computer language called C++, it looks like an exclamation point followed by an equal sign: !=.

EXAMPLES

28 != 28 28 "is not equal to" 28	*False*
0 != 7 0 "is not equal to" 7	*True*

EXERCISES

Evaluate each relational expression as *true* or *false* given the values contained in the variables.

x	*y*	*answer*	*first_val*
14.6	–15	24	10

1. $x < y$
2. *answer* >= *x*
3. *first_val* < *answer*
4. $x >= y$
5. $y = = -15$

3.6 LOGIC OPERATORS

The logic operators are a group of operators that allow the computer to make a complex decision. Let's start with examining the logic involved in some statements that you encounter in everyday life. Then we will take what we have learned and apply it to our context of programming in a language.

LOGIC OPERATORS IN THE REAL WORLD

Imagine a situation where you are only allowed to go to a party if exactly one of the following numbered choices is fulfilled:

You can go to the party if

1. you take out the garbage *and* you clean your room
2. you take out the garbage *or* you clean your room
3. you rake the leaves *or* you wash the floor
4. you rake the leaves *and* you wash the floor

Which of these numbered conditions seems the most appealing? If you think of each of these situations *logically*, #2 and #3 are the easiest to fulfill. This is not because of the simplicity of the tasks involved. (Let's ignore the fact that you would do anything to avoid having to wash the floor.)

Choices #2 and #3 are easiest because of what is meant by the word "or" linking the two tasks. By using this word, the underlying assumption is that you have a *choice* in what to do in order to gain permission to go to the party. Choices #1 and #4 give you no choice: You must do *both* tasks before getting permission. Look at the table and examine all the possibilities of doing either, both, or neither of the tasks and the resulting permission.

A check means that a task is completed and an X means that a task is not done.

Tasks		*Permission*
You take out the garbage *and* you clean your room		
√	X	NO
You take out the garbage *and* you clean your room		
X	√	NO
You take out the garbage *and* you clean your room		
√	√	YES
You take out the garbage *and* you clean your room		
X	X	NO
You rake the leaves *or* you wash the floor		
√	X	YES
You rake the leaves *or* you wash the floor		
X	√	YES
You rake the leaves *or* you wash the floor		
X	X	NO
You rake the leaves *or* you wash the floor		
√	√	YES

Look at each situation to see why permission was either given or denied. What can we say about the "and" conjunction? It requires *both* conditions to be executed. In contrast, the "or" conjunction needs only one condition to be checked to get permission. Each of these conjunctions is called a *logic operator*. The "and" conjunction has been used with one task on either side of it; therefore, it behaves like a binary operator. Likewise, the "or" conjunction has been used with one task on either side of it and so is also a binary operator. Both words are called *logic* operators because they evaluate a statement logically just like you would in everyday circumstances.

EXERCISES

Under each task, mark one check (√) or both so that permission will be granted or access will be allowed for each situation. For some problems, more than one answer is correct.

1. You can gain access to the account if

   you have a magnetic strip card *and* you know the password.

   ______ ______

2. You can gain access to the safe deposit box if

   you have a key *or* you have a photo ID.

   ______ ______

3. You can go to the concert if

   you order the ticket over the phone *and* you pick it up before 6 p.m.

   ______ ______

4. You can eat dessert if

   you ran 2 miles *or* you walked 3 miles.

   ______ ______

5. You can see the movie if

   it is rated PG *or* it has the word "Lassie" in the title.

   ______ ______

LOGIC OPERATORS IN THE COMPUTER LANGUAGE ENVIRONMENT

In the context of computer programming, we do not deal with tasks but *true/false results* that are generated by *relational expressions*. (Recall that a relational expression uses a relational operator for comparison—such as greater than, less than, not equal to, etc.) Each task in the real world is a relational expression in the computer language environment. Let's compare one of our previous examples with an example appropriate in a programming environment.

You take out the garbage *and* you clean your room	Real world example
x must be greater than 24 *and* x must be less than 30, which looks like the following: $x > 24$ *and* $x < 30$	Programming example

You rake the leaves *or* you wash the floor	Real world example
y is less than 0 *or* *y* is greater than 100	Programming example
$y < 0$ *or* $y > 100$	

Recall that if both tasks are done (in the case of the "and" operator), then permission is granted or access given. *In the programming example, a relational expression* will appear on both sides of the operator. Depending on whether one or both of these expressions are true, the result will be true or false.

The logic operators are these words: *and* and *or*. Like any binary operators they must have an *operand* on either side. The *operand* on each side *must be* an expression that has a *true/false* value: So far, only the *relational expression* gives such a value. *Arithmetic expressions* yield numbers and are not acceptable as operands for the logic operators.

When you evaluate a logic expression, keep in mind two things: Once there is a *false* expression on one side of an "and" operator, the *whole statement is false*. Once there is a *true* expression on one side of the "or" operator, the *whole statement is true*.

EXAMPLES WITH VARIABLES

sum > 12 *or* *sum* < 0

What will be the result if *sum* contains 14?
What will be the result if *sum* contains –25?
What will be the result if *sum* contains 12?

In the first example, 14 > 12 so we have one *true* result in an *or* statement. Therefore, the entire statement is *true* regardless of the second relational expression's result.

In the second example, –25 is not > 12 but –25 < 0 so a *false or true* result is overall *true*.

In the last example, 12 is not > 12 and 12 is not < 0 so *false or false* produces *false*.

EXAMPLE

answer > –5 and *answer* < 20

What will be the result if *answer* contains 10?
What will be the result if *answer* contains –65?
What will be the result if *answer* contains 28?

In the first example, 10 > –5 and 10 < 20 so we have a *true* result on either side of the "and." Therefore, the entire statement is *true*. In the second example, –65 is not > –5, although –65 is < 20, but the entire statement is *false* because we have a false on one side of the "and." In the last example, 28 is > –5, but 28 is not < 20 so *true and false* produces *false*.

HINT!

The logic operator *and* has to have all relational expressions around it to be true for it to generate an overall *true* answer. The logic operator *or* needs *only one* of its operands to be true to generate an overall *true* answer.

HINT!

The *and* operator is usually the word *and* or double ampersands: &&. The *or* operator is the word *or* or it is represented symbolically as two vertical bars: ||.

EXAMPLES

x > 3 *and* x < 10	is the same as	x > 3 && x < 10
y > 100 *or* y < 0	is the same as	y > 100 | | y < 0

EXERCISES

Evaluate whether each is *true* or *false* given the values assigned to the variables as follows:

sum = 15; *answer* = –25; *x* = 52.6; *y* = 14.75;

(We will use the symbols we mentioned earlier: && for "and," | | for "or.")

1. *sum* > 10 && *sum* < 100
2. *answer* < –100 | | *answer* > –50

3. $x < 53$ || $x >= 0$
4. $y < 20$ && $y > 15$

3.7 A SPECIAL LOGIC OPERATOR: THE NOT OPERATOR

Our last logic operator is the word *not*. What does a *not* operator do? To understand the *not* operator, let's look at an example from the real world. Imagine that you have a very contrary friend (lucky you!). She changes everything you say. Let's look at some of your conversations.

You: That movie was great!
Her: No, it was horrible!

You: I liked the opening scene.
Her: No, you didn't. You said it was lame.

You: I didn't like the ending, however.
Her: Yes, you did. You were laughing all during it.

After a series of conversations like these, you might want to avoid your friend for a while. Your friend behaves like the *not* operator. Everything that you have said gets changed, *logically*. If you said you liked something, the not operator (your friend) will say you didn't like it. If you say you didn't like something, the not operator (your friend) will say you liked it by altering what you said to its *logical opposite*. Let's take a few sentences and apply the *not* operator to them and see what doing this does to the meaning of those statements. Notice that the not operator *precedes* the statement that will be altered.

Operator	Statement	Resulting Statement
Not	(It's raining outside.)	*It's not* raining outside.
Not	(I have no homework.)	I *have* homework!
Not	(I disliked the movie " Gladiator.")	I *liked* the movie "Gladiator."

The *not* in front of each of these statements changes the logic of whatever it is applied to.

It changes every statement or expression it operates on. It is a *unary* operator because it only needs *one operand* on which to work.

In the next set of examples, we will look at a resulting statement and determine how a not statement was used to create it.

Resulting Statement	not	Original Statement
I love the summer.	not	(I don't like summer.)
He reads a lot.	not	(He doesn't read much.)
She did not write a letter.	not	(She did write a letter.)

EXERCISES

Write the resulting statement from each *not* statement given.

1. not (I like ice cream.)
2. not (She left before 2:00 p.m.)
3. not (They didn't take the train.)

For each resulting statement, write a *not* statement that could have produced it.

4. I like listening to country music.
5. I don't remember his last name.
6. She lives in Riverdale.

HINT!

The *not* operator could be the word *not* or the exclamation point, !.

EXAMPLES

not (x < 3) is the same as ! (x < 3)
not (y > 100) is the same as ! (y > 100)

WATCHOUT!

The *not* operator does not always produce a negative statement; it produces the *opposite* of its operand.

Consider these examples with variables. In each example, a variable is first assigned a value and then a *not* expression is used. Follow each example to see what the result is for each.

Expression	Result
sum = 14;	
! (*sum* > 12)	
14 > 12	
⇓	
true	
not true	false

Because 14 *is* greater than 12, the original statement (*sum* > 12) is *true*. However, the use of the *not* operator on the result, *true*, has the effect of changing the overall expression to the value, *false*. To be "not true" is to be "false."

answer = –60;	
! (*answer* > = 78)	
–60 > = 78	
⇓	
false	
not false	true

In this example, the original statement, "negative 60 is greater than or equal to 78" is *false*. By using the *not* operator, however, the value of the entire expression becomes *true*. To be "not false" is to be "true."

first_ans = 0;	
! (*first_ans* = = 0)	
0 = = 0	
⇓	
true	
not true	false

In this example, the relational expression inside the parentheses is done first. Because 0 is equal to 0, the relational expression is *true*. After that, the *not* operator changes the value from *true* to *false*.

WATCHOUT!

The *not* operator is always used with parentheses as you have seen in the previous examples. Why is this so? It is important to group together what it is that is being *altered*. By using parentheses around the expression, the computer is being instructed to find the value of the expression in parentheses (P.E.M.D.A.S.) *first* before altering the value by applying the *not* operator.

In the following examples, we will use two relational expressions with a logic operator outside the parentheses. See how these work.

```
Expression                          Result
        y = 36;
! (14.5 < y || y > 39)
: 14.5 < 36 || 36 > 39
:     ⇓         ⇓
:   (true || false)
:         ⇓
:    true or false
:         ⇓
:       true
not     true                        false
```

Because parentheses are to be done first, we evaluate the relational expressions within. The first relational expression (14.5 < 36) is *true* and the second (36 > 39) is *false*. Because the operator "or" is used between them—*true or false* gives us the value *true*. (Notice that we wait to use the *not* operator because we are still inside of the parentheses.) In the last step, the *not* operator changes the value from *true* to *false*.

```
Expression                          Result
        val = 35;
! (val > 23  &&  val < 30 )
:  35 > 23   &&  35 < 30
:     ⇓             ⇓
:    true    &&   false
```

```
        ⇓            ⇓
      true   and   false
              ⇓
            false
not         false                 true
```

Again, we evaluate what is inside the parentheses first. The first relational expression (35 > 23) is *true* and the second relational expression (35 < 30) is false. However, because the operator "and" is used between them, true *and* false yields false. The last step with the *not* operator produces true because "not false" is "true."

EXERCISES

Given the values in each of these variables, evaluate each statement as *true* or *false*.

sum	*result*	*c*	*my_ans*	*last_val*
–67	23	8.6	12.5	31

(Recall: We will use the symbols: &&, | |, and ! to represent "and," "or," and "not.")

1. *sum* < 14 | | *sum* > 25
2. *result* >= *my_ans* && *result* < 23
3. ! (*c* > *my_ans*)
4. ! (*last_val* < *c* | | *c* > 0)
5. –100 < *sum* && *sum* < –45

3.8 A POWERFUL OPERATOR FOR ANY COMPUTER LANGUAGE: MOD

In addition to each of the arithmetic operators already mentioned, most languages provide an additional operator in division. It is called the *mod* operator, short for *modulus* in Latin, or *remainder*. To understand what it does, let us revisit division between two integers. The *mod* operator, when used between two operands, produces the *remainder* of a long division problem between the two integer operands. The *mod* operator is usually represented by the % symbol or the word *mod*. In the next section (3.9), we will show you how it is used in programming.

28 mod 14 is 0 because there is no remainder
172 mod 35 is 32 because 172/35 = 4 with a remainder of 32
1943 mod 7 is 4 because 1943/7 = 277 with a remainder of 4
18 mod 17 is 1 because 18/17 = 1 with a remainder of 1

In each case the number to be divided is of greater value than the one it is being divided by (the divisor)—(e.g., 1943 is being divided by 7. In all of these cases there has to be a remainder (even when it is 0; Figure 3.2).

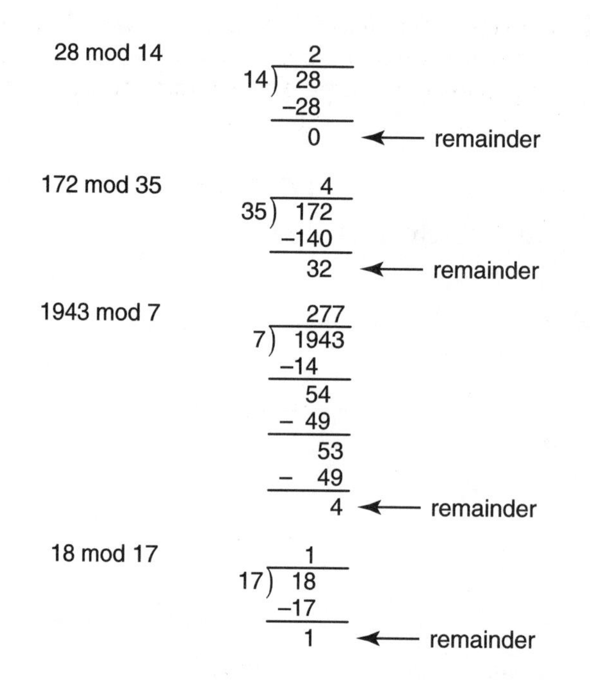

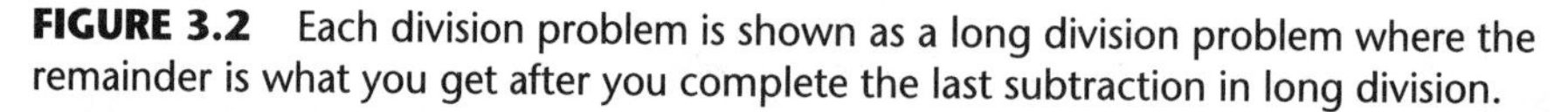
FIGURE 3.2 Each division problem is shown as a long division problem where the remainder is what you get after you complete the last subtraction in long division.

Now consider some interesting examples where the *divisor*, the number by which you divide, is greater than the *dividend*, the number under the long division symbol. Also notice what happens when you use negatives with the mod operator.

38 mod 47 is 38
Remember that the mod operator produces the remainder so 38/47 = 0 with a remainder of 38.

–14 mod 5 is –4

When we divide –14 by 5 we get –2 with a remainder of –4.

–32 mod –6 is –2

We see that –32 divided by –6 is 5 with a remainder that is negative (–2).

WATCHOUT!

It is very easy to make mistakes with the mod using negatives. The answers you get always come from the arithmetic of a long division problem—and not from the rules from algebra regarding negatives and positives.

If you ever have an example using the mod operator and you do not understand why the answer is negative, revisit a long division problem to see how the negative answer is generated (Figure 3.3).

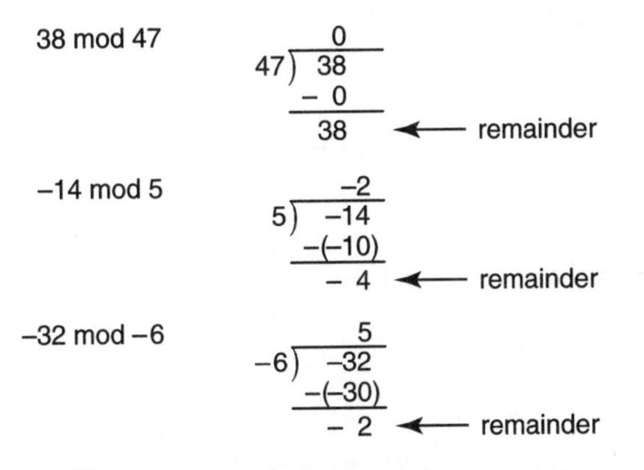

FIGURE 3.3 In each example, the long division shows the remainder that is produced. When using the mod operator and a negative integer, note whether the remainder is positive or negative.

HINT!

The *mod* operator is *used only between two integers*. It is not designed for use with real numbers (numbers with decimals, etc.). An *error results* from trying to use the *mod* operator with anything other than integers.

EXERCISES

Evaluate each mod expression given the values in the variables. (Recall: We will use the symbol for the mod operator, %.)

sum	*result*	*my_ans*	*last_val*
14	6	19	–3

1. *sum % result*
2. *my_ans* % 5
3. *result % last_val*
4. *my_ans % sum*

3.9 HOW IS THE MOD OPERATOR USED IN PROGRAMMING?

This operator can be used in very interesting ways. By being able to tell whether there is a remainder from doing a division problem between two numbers, you can tell whether one number fits *exactly* into another number. Why would this be useful? Look at these questions that the *mod* operator can answer if used appropriately.

1. *Do we have a divisor of another number?*
   Is 35 a divisor of 70? Yes, because 70 % 35 is 0. If the result of using the *mod* operator between two integers is *zero,* then the right-hand operand (35) is a *divisor* of the left-hand operand (70). Therefore, 35 is a divisor of 70 because 35 fits perfectly into 70, with no remainder (i.e. with zero remainder).
2. *Do we have an even number?*
   An *even* number is a number divisible by 2. Let's say that you have an unknown number contained in the variable *x*. If *x* % 2 is 0 then *x* is an *even* number. Some examples: 46 % 2 is 0, 8 % 2 is 0.
3. *Do we have an odd number?*
   Similarly, if an unknown number contained in *y* is used in a mod statement, you can determine whether *y* is an odd number. If *y* % 2 is 1 then you have an odd number.

Some examples: 13 % 2 is 1. 25 % 2 is 1.

SUMMARY

We defined *operators* as the *actions* taken on numbers or variables. The *precedence* of an operator or its *priority* in terms of when it should be executed was introduced. The terms *binary* and *unary* operators were defined in terms of the *number of operands* required for an operator to function properly.

Next, different kinds of operators were defined: *arithmetic, relational,* and *logic* operators. The *arithmetic* operators are the most familiar because they involve the operations of *addition, subtraction, multiplication,* and *division*. There is a special case in division—*division between two integers*—where any *fractional part* in the answer will be *dropped*. The *relational* operators (<, <=, >, >=, ==, ! =) produce *true/false* answers as do the *logic* operators (&&, | |, !).

Finally, the *mod operator* (%) was defined and some instances of its use in programming were given. In the next chapter, we will begin to look at some short programs using what we have learned from Chapters 2 and 3.

ANSWERS TO ODD-NUMBERED EXERCISES

3.2

1. c 3. b 5. 3 (2 –5) + 6 * 6 7. –16 9. –41
2 1 4 3

3.3

1. binary 3. unary, binary

3.4

1. 1 3. 27 5. 25.79 7. 1.67

3.5

1. false 3. true 5. true

3.6 (i)

1. √, √ 3. √, √ 5. √, X or X, √ (two possible answers)

3.6 (ii)

1. true 3. true

3.7 (i)

1. I don't like ice cream. 3. They did take the train. 5. not (I do remember his last name.)

3.7 (ii)

1. true 3. true 5. true

3.8

1. 2 3. 0

CHAPTER

4

Programming: It's Now or Never!

IN THIS CHAPTER

- Review of Declaration and Assignment
- Writing an Output Statement
- Understanding the *Cout* Stream
- The *Endl* Command
- How to Insert Comments into a Program
- Introduction of *Compiler Directives*
- The Main Section of a Program
- The *Return* Statement
- Three Short Programs

4.1 PUTTING A PROGRAM TOGETHER

We are almost ready to write some short programs. In the past two chapters, we have looked at some of the initial aspects of programming—

namely, declaring and assigning variables, and manipulating those variables through operators. Throughout this chapter and the following ones, we will use the C++ programming language commands.

DECLARE, ASSIGN, AND MANIPULATE

If you recall, when *declaring* a variable, we tell the computer that a variable is of a certain *type* so that the computer can set aside the appropriate amount of memory. In the programming language C++, the word for integer is shortened to *int*. Let's declare two integer variables as follows:

int *first_val, second_val;*

Now we have told the computer that two variables called *first_val* and *second_val* will be used in our program. We have just *declared* the variables (i.e., introduced them) to the computer. The next step is to *assign* the variables. We will let the programmer assign the *first_val* variable. (Let's leave the second variable, *second_val,* alone for a moment.) The following statement will accomplish *first_val's* assignment:

first_val = 25;

Our third part of this process involves using one of our operators from Chapter 3. We will use an arithmetic operator, the multiplication sign, '*.' Here we will also use the second variable to hold twice the contents of *first_val's* value.

second_val = 2 * *first_val;*

This means that *second_val* has the value of 50 because 2 * *first_val* (25) is 50.

TIME FOR AN OUTPUT STATEMENT

In our next stage of writing a program, we should show the contents of *second_val* on the screen so that our user will know we doubled the value of *first_val* and produced *second_val* as a result. We want the user to be able to see his *output,* the data that the computer has generated through the manipulations we performed. An *output statement* is a programming *language statement* that generates *output*—data contained within variables or messages to the user *to be directed to* the screen.

An output statement in the programming language of C++ will have three key parts: a *stream name, an insertion operator,* and *a variable* or *message.* The stream name is the name of a *channel* where data is sent *before* it goes onto the screen. An *insertion operator* is an operator that inserts data into that stream. After the insertion operator, either a variable or a message can be placed. The variable's *value* will be inserted into the stream and then shown on the screen.

Because the programming language C++ is a high level language, you do not have to worry about how the stream sends its data to the screen. That does not concern us. We just need to ensure that we are sending all data and messages properly to the stream.

COUT

Cout pronounced "see-out" is the name of the *stream* where data gets sent before it is channeled to the screen. (This stream is the opposite of the "cin" stream we introduced in Chapter 3.) When you use a "cout" statement, you are really sending data to the screen through the *cout* channel. Anything that is sent to this stream will eventually end up on the screen where it can be viewed.

The syntax for using the *cout* stream goes as follows. You name the stream, *cout,* then you follow with the *insertion operator* (this symbol : <<), which indicates that something is going *out into* the stream. Next, you put the variable name or message that you wish sent to the screen (via the stream). Messages must be in quotation marks. Let's look at some examples.

EXAMPLES

```
cout << first_val;          where
cout << " hello.";          (first_val = 25)
```

In the first example, we are sending a variable to the stream which "feeds into" the screen—that really means that the contents of the variable will appear on the screen. In the second example, a message is sent to the stream and will be displayed on the screen next to the number in *first_val*. It will look like this:

```
25 hello.
```

When you wish to put *more than one item* into the stream you need to use the insertion operator *before* each item. Let's look at some other examples where we use the *cout* stream. In both examples that follow, more than one item is sent to the stream so the insertion operator is placed *in front of each* variable or message.

cout << "Here is the first value:" <<*first_val;*

produces:

Here is the first value: 25

cout <<*first_value* << *second_val*; (where Second_val = 2* first_val

produces:

25 50

In our third set of examples, let's say you wish to put a period, "." to finish a sentence that included a variable. You would need to use the insertion operator *between* the variable and the period because the period is a string following a variable.

cout << "The content's of second_val is"<< second_val << ".";

WATCHOUT!

Any output should follow the *cout* command and the insertion operator, << ; if there is more than one item of output, i.e., multiple variables or messages, put the insertion operator *before* each item.

Now let's say that you were using the *string* type mentioned from Chapter 2. Recall that a string can hold *several* characters or words. The string type could hold a message if you assign the message to the string. Let's declare and assign a message to a string type.

```
string   my_message;
my_message = "Have a nice day.";
```

Now we can send the message to the stream without using any quotation marks because we will use the variable, *my_message*, to hold the sentence.

```
cout << my_message;
```

HINT!

Remember that variables do not need quotation marks; it is assumed that when a variable is sent to the cout stream, its *value* will be sent to the stream and, ultimately, to the screen.

HOW TO EXECUTE A LINE FEED: ENDL

The *endl*, pronounced "end line," command (in the programming language C++) causes a line feed on the screen. To have data and messages appear on separate lines, you need to direct a line feed to the screen. The way to do this in the C++ programming language is to put the *endl* command *into* the *cout* stream. When the data "flows" from the stream onto the screen, the *endl* command will cause a line feed so that anything *that follows* this command will be on a new line. It is very important to notice that if an *endl* is used at the end of a *cout* statement, the *next cout* statement will show data on the *next line.*

Let's look at some examples *with* and *without* the *endl* command:

```
cout << first_val;
cout << "Goodbye.";
```

```
screen
25  Goodbye.
```

```
cout << first_val << endl;
cout << "Goodbye.";
```

```
25
Goodbye.
```

```
cout << "Hello " << endl << endl << "Goodbye.";
```

```
Hello

Goodbye
```

Notice that there were two line feeds—so that "Goodbye" is on the line *after* the next line.

COMMENTS

Another useful tool in programming languages is the ability to *comment* in a program or *put descriptive remarks* next to a programming statement or statements. The reason for putting comments into a program is to make the code *clearer* for any reader of the program.

Look at this example from everyday life:

Sentence	*Comment*
Jack will be late today.	Jack is the man who used to work with Marie.

Programming Statement	*Comment*
`first_val = 25;`	`//Programmer is assigning first_val.`

Notice the symbol, //, used to the left of the comment. When you write a comment for a program, you do not want the compiler to think that your comment is part of the *code that is being executed.* Before every comment you write, the symbol, "//," tells the compiler to *ignore* what follows on that line.

If your comment is *longer than one line,* you need to use two different symbols—one to indicate the beginning of the block to be ignored (/*) and then another to indicate the end of the block (*/). These symbols function like parentheses. After the second symbol is read, the compiler "knows" it can start *executing* code again.

```
/*  Everything to the right of these symbols is ignored.
Write whatever you want in here
since this code will be ignored.
The end of the block is to the right. -> */
```

4.2 COMPILER DIRECTIVES

Now that we have seen four parts of early program writing—declaring variables, assigning them, manipulating their values, and displaying output—what else does a program need?

Compiler directives sounds like a complicated term but it is not. The first thing is to recall what the *compiler* does. The *compiler* translates our high level language code into low level and, ultimately, machine language code that the computer understands.

Directives is just a fancy word for *directions.* So *compiler directives are special directions for the compiler.* Although there might be several compiler directives, we are only interested in a specific directive, *the include directive.*

THE INCLUDE DIRECTIVE

The *include directive is a special instruction* (in the C++ programming language) for the compiler *to get a specific file* that we ask for *and insert it at the top of our program.* That way, our program can benefit from any of the capabilities that the included file offers. When the compiler starts to translate our program into low level code, it first gets the file that was mentioned at the top of the program and starts to translate that code. Whatever that file *can* do, our program can now *benefit* from it.

There are several files that we might like to use in any program that we write in the programming language C++. These files have names that all end in the *extension*, ".h," which stands for the classification of a *header* file. An extension is an appendix used to indicate what type of file you have. (You may have seen other extensions that are attached to file names e.g., ".jpg" and ".gif" which refer to files that comprise pictures.) A *header* file is a file that can be placed at the *top* of a program and accomplishes certain tasks that the file including it needs. (Later, in Chapter 16, we will discuss header files in more depth.) Here are some examples of header files we might like to use at some point in our programming:

Header File Name	Description
iostream.h	manages the streams used for input and output of data
string.h	manages the string type variable
math.h	allows many math functions to be performed on data—similar to those on a calculator (sin(x), abs(x), etc.)

If you recall, *input* is necessary when someone other than the programmer wishes to load data into variables. That is, the user wishes to send values into variables. *Output*, or sending data to the screen, is a basic aspect of most programs. There are very few programs that would not send some results to the screen to be viewed. For these reasons, practically all programs need access to the *streams that channel data to and from the keyboard and screen*. The "iostream.h" (pronounced "eye-oh-stream dot h") file allows us to use both the *cout* stream for displaying output on the screen and the *cin* stream (from Chapter 2), which allows the user to assign variables through keyboard input. These two streams are part of this file.

At the top of our program (written in the programming language C++), we must give the compiler directive to get "iostream.h." The directive looks like this:

```
#include < iostream.h>
```

The number symbol, #, indicates that this is a *directive* to the compiler. The *iostream.h* is the *name* of the file that manages input and output. In the chapter on classes in C++ (Chapter 16), you will learn more about these header files.

HINT!

We will put *# include < iostream.h>* at the top of *every* program we write in C++ because we always expect to have input and output in our programs.

4.3 THE MAIN SECTION OF A PROGRAM

Programs are broken into sections. At the top of a program are any files that might help our program accomplish its task. After that, a program can be broken into sections where each section accomplishes a specific task. The reason for the subsections is really one of *organization*. By blocking code into separate sections, you are organizing the code so that any reader can understand or fix it, if necessary. (This is especially important if there are *errors* in the language code.)

THE MAIN SECTION

The *main section* contains the *body* of a program. With the first programs we write, it is not necessary for us to move away from the main to other subsections. All the code that we write to execute some task will be contained in this main section. Later we will learn how to compartmentalize a large program—that is, break it into sections that each do some task rather than having all the code in "the main."

The main section has a *heading* (like a title) and is followed by two *braces* : an opening brace ({)and a closing brace (}) to indicate where the main section both begins and ends. Inside the braces go the programming statements that you write.

```
int main ( ) // the heading of the main section
{            // the opening brace

//***Your programming statements go between these braces.***

return 0;    //the return statement
}            //the closing brace
```

In the heading, you see the word for an integer, *int,* followed by the word "main" and then some empty parentheses followed by the braces that begin and end the main section.

In Chapter 7, we will explain the syntax of this heading. Just use it for now.

THE RETURN STATEMENT

The return statement is the last statement of the main section and you might consider the *return* statement in the following way. Imagine that the compiler has been given a key to the main room—the control room of the program. This room is usually locked because it is the control center and we don't want just anyone going in there. When the compiler is done with the main section, it "returns" the key to the room—for security reasons. This "key" is an integer according to the first word in the heading of the main. The compiler is being directed to *return an integer* before it leaves the main section.

Because many programs don't necessarily produce an integer, we come up with the idea of returning the integer 0 as a matter of simplicity. By using the *return 0* ("return zero") command at the bottom of the main section, the programmer is satisfying the compiler's requirements to generate an integer before leaving the main section and closing the program for good. Once it does that, the program is over and the compiler's work is finished. In Chapter 7, we will learn more about how this statement works, but for now, this explanation should suffice.

```
int main ( )
{

//Your programming statements go between these braces.

return 0;
}
```

The heading gives some information about how the main must function. An integer must be produced before we can "close" the main. The last command, *return 0,* allows the compiler to *leave* the main section carrying the integer 0 and "know" that it has finished its work there.

4.4 BUILDING A PROGRAM OUTLINE

We are now ready to build an *outline* of a program. The first part of the program should include any directives to the compiler. The next part will be the main section blocked off by the opening and closing braces: { }. So now our outline looks like this:

```
#include <iostream.h>   // the compiler directive

int main ( )        //  the heading
{

// Your programming statements go between these braces.

return 0;  //returning an integer so that we can satisfy
//the heading's requirement of an integer being produced.
}
```

Now let's look at some of the code (programming statements) we talked about previously. We mentioned that it was necessary to include the declaration of a variable. The program outline can be completed in this way:

```
#include <iostream.h>
int main ( )
{
  int first_val, second_val; // the declaration section


return 0;
}
```

Next, we can insert the code that causes assignment of values to variables.

```
#include <iostream.h>
int main ( )
{
  int first_val, second_val;

first_val = 25;      // the variable is assigned
second_val = 2 * first_val; // second_val assigned with
                            // twice first_val's value

return 0;
}
```

In the next example we include the output statement.

```
#include <iostream.h>
int main ( )
{
  int first_val, second_val; // the declaration section

first_val = 25;      // the variable is assigned
second_val = 2 * first_val; // second_val assigned with
                            // twice first_val's value.
cout << second_val << endl; // second_val's value is printed.
return 0; //Execution of the main section will now end.
}
```

4.5 SOME SHORT PROGRAMS

Now that we have examined the outline of a program, we are ready to see some sample programs. In each program we write, look at the basic structure of the program and keep in mind that it will probably contain three to four elements:

1. Declaration of variables.
2. Assignment of variables.
3. Manipulation of those variables.
4. Variables' values printed on the screen.

EXAMPLES

I. A program that computes the average of three numbers.

Description
We will declare three variables to hold three distinct real numbers (assigned by the programmer) and then we will compute the average of those numbers and display it on the screen. In addition to declaring three doubles to hold three real (non-integer) numbers, we need to declare a variable to hold the average; we'll call that variable, *average*.

Notice the use of the two arithmetic operators, + and / to compute the average. Because we are using doubles (non-integers) with the division symbol, the compiler will perform division just as a calculator would by displaying the remainder in decimal form.

After examining the code in the following program, look at the output produced if the program is run on a computer. Recall that output is anything displayed on the screen—the data values that the variables contain after those variables have been manipulated.

The Program

```
#include <iostream.h>
int main ( )
{
double  first_val, second_val, third_val;
                                  //declaration section

first_val = 25;      //programmer has assigned all
second_val = 38.9;  // three variables.
third_val = 42.7
average = ( first_val + second_val + third_val) /3;
cout << "The average of the three numbers is "<< average << " ."
<< endl; // Note how this line wrapped to the next line.

return 0;
}
```

```
Output

The average of the three numbers is   35.53 .
```

II. A program that computes the average of three numbers from the user.

Description

Let's take that same program and let *the user assign* the variables. Remember that we should first ask the user for the values so that he will understand what to do.

The Program

```
#include <iostream.h>
int main ( )
{
double  first_val, second_val, third_val;
                                  //declaration section
cout << "Please type three numbers." << endl;

cin >> first_val >> second_val >> third_val;//the user is
//assigning variables by typing numbers at the keyboard.

average = ( first_val + second_val + third_val) /3;
cout << "The average of the three numbers is " << average << " ."
<< endl; // Note how this line wrapped to the next line.
return 0;
```

```
}
```

```
Output

Please type three numbers.
25  38.9  42.7
The average of the three numbers is   35.53 .
```

Of course, the value shown will depend on what the user types.

III. A program that prints the user's name on the screen.

Description

In this example, we will ask the user for his name and store that value in a string. Then we will display the string's value on the screen.

The Program

```
#include <iostream.h>
#include <string.h> // Here is the second directive to the
//compiler.
int main ( )
{
string  name;  // declaration section
cout << "what is your name?" << endl;
cin >> name;//User assigns value by typing it at the keyboard.
cout << "Your name is "<< name << " .";
return 0;
}
```

```
Output

What is your name?
Jack
Your name is Jack .
```

IV. A program that updates a checking account balance by recording an amount as a debitæsomething to be deducted from an account balance.

Description

In this example, let's ask the user for the amount he wishes to subtract from his checking account balance. Then we will subtract that amount and display the new balance.

The Program

```
#include <iostream.h>
int main ( )
{ double balance, amount; //declaration section
cout << "What is your balance?" << endl;
cin >> balance;
cout << "What is the amount of your check?" << endl;
cin >> amount;
balance = balance - amount;
cout << "Your balance is now " << '$' << balance << '.' << endl;
return 0;
}
```

```
Output

What is your balance?
345.67
What is the amount of your check?
26.75
Your balance is now $318.92.
```

Although these programs are limited in what they do, they give you an idea of how a program is structured. You need to remember that all variables should be *declared prior to use,* then *assigned by either the programmer or user.* After that you can *manipulate the variables* and *display output.*

In the next chapter, we will learn to make decisions on the computer—this will allow us to do a lot more interesting programs. Some examples of programs that involve decisions are choosing the largest of a group of numbers or giving the user a choice in responding to a menu of choices. None of these possibilities can be accomplished without some sort of *decision-making ability* in a programming language.

EXERCISE

1. Write a short program that asks the user for his name, age, and grade level and then prints that data on the screen.
2. Write a short program that will take two numbers determined by the user and do the following with them: compute their sum, difference, product, and quotient. (Recall that a product results from multiplying two numbers and a quotient results from dividing two numbers.)

SUMMARY

In this chapter, we looked at the essential elements of a program: declaration, assignment, and manipulation. To be effective, most programs will display their results on the screen: What is displayed on the screen is known as *output*. The use of the *cout* stream and the insertion operator (<<) allow us to produce output. The *endl* command allows us to execute line feeds on the screen.

Another good feature of clear programming is the careful use of comments. *Comments* can be used to explain the purpose of programming statements to any reader of a program. When you put a comment in a program, you need to use some symbol so that the compiler "knows" it is not executing a command. In C++, this symbol is the one line comment "//" or for multiple line comments, the symbols, "/*" and "*/," to begin and end the section.

Before writing a program in the programming language C++, we need to learn a few useful parts that comprise all C++ programs. The first is to *include a directive* to the compiler regarding input and output. To be able to display messages on the screen or get data from the user, you need to include the *iostream.h* header file. This file allows us to display messages and take in data.

Another part of all C++ programs is the *main section*. The *main* is the body of the program. The *heading* of the main requires that an integer be generated before the main section is completed. Opening and closing braces, { }, show us where the main begins and ends. Once the main has been completely executed, the statement *return 0* causes the compiler to exit the main properly—thereby ending the program.

In the last part of the chapter, there are some examples tying all these elements together into three programs. In the next chapter, we will learn how to control decision making so that we can expand our programming capabilities.

ANSWERS TO EXERCISES

1.

```
#include <iostream.h>
#include <string.h>
int main ( )
{
```

```
str  name;
int age, grade_level;
cout << "What is your name?" << endl;
cin >> name;
cout << "What is your age?" << endl;
cin >> age;
cout << "What is your grade_level?"<< endl;
cin >> grade_level;
cout << name << "  "<< age << "   " << grade_level << endl;
return 0;
}
```

2.

```
#include <iostream.h>
int main ( )
{
double  first_val, second_val;
double  sum, prod, diff, quo;

cout << "Please type two numbers." << endl;

cin >> first_val >> second_val;

sum = first_val + second_val;
prod = first_val * second_val;
diff = first_val - second_val;
quo = first_val / second_val;

cout << "The sum is " << sum << "." << endl;
cout << "The product is " << prod << "." << endl;
cout << "The difference is " << diff << "." << endl;
cout << "The quotient is " << quo << "." << endl;

return 0;
```

CHAPTER

5

True or False: The Basis for All Decisions

IN THIS CHAPTER

- The Boolean Type
- Developing a Model of a Decision
- Program Flow
- Control Statements
- The If...Statement
- The If...Else...Statement
- Switch/Case Statement

In Chapter 3, we examined different kinds of operators, namely the arithmetic and relational operators, as well as the logic operators. The arithmetic operators produce a number. The relational and logic operators produce a *true/false* result. They require a different type variable for holding this result; it is called the *boolean* type. We will use the *boolean* type in the decisions that we learn to program on the computer.

5.1 THE BOOLEAN TYPE

In addition to the types we have already studied—the integer, double, character, and string types—there is another type, the *boolean*, named after the mathematician George Boole who did extensive work in the subject of logic. It is a variable type that holds the result *true* or *false*. These values are not the strings "true" or "false." They are the *values* true or false. The only way to get these values is to use the relational operators (<, >, etc.) and/or the logic operators (and, or, not) from Chapter 3. When you compare two numbers with a relational operator, you get a value of *true* or *false*.

Example	*Result*
16 > 15	*true*
8.5 < = 8.2	*false*
–12 < 4	*true*

Now let's declare some boolean variables and use them in assignment statements with these examples. Because we want to declare *more than one* boolean variable at once, we will make *a list* of variables after the type boolean. Just separate the variables by commas before ending the declaration statement with the semicolon. Let's use the name "flag" for one of the boolean variables because flag reminds us of the expression to "raise a flag" and catches our attention. You will later see why this is useful.

variable type variable, variable, etc. ;

```
boolean flag, x, answer;        //flag, x and answer are all
                                //boolean type variables
```

Example	*Result*
`flag = 16 > 15;`	flag holds *true*
`x = 8.5 <= 8.2;`	x holds *false*
`answer = -12 < 4;`	answer holds *true*

Now let's consider some examples where the boolean type is used with some *variables* in relational expressions. In the previous examples, the values were used directly in the relational expressions.

```
int  a, b, c ;//declaring three integers in a list
boolean answer, flag, result; //declaring three booleans
a = 14;
```

```
b = 0;
c = 7;
```

Example	*Result*
`answer = a < c ;`	answer holds *false* because 14 is *not* less than 7 14 < 7
`flag   = b > c;`	flag holds *false* because 0 is *not* greater than 7 0 > 7
`result = a > b;`	result holds *true* because 14 is greater than 0 14 > 0

EXERCISES

Give the value of each boolean variable for each statement. *Answer*, *flag* are declared as booleans.

1. answer = 14 < 6;
2. flag = 10 >= 10;
3. answer = 18.1 < 18.01;
4. flag = –28 < –35;

Given the values of the integers, evaluate each boolean as true or false.

int x, y, z;
x = 42; y = 25; z = –6;

5. *answer* = x > y;
6. *flag* = y > z;
7. *answer* = x <= z;
8. *flag* = y < x;

HINT!

When you want to declare more than one variable of the same type, declare the type and then follow with the list of variable names, separated by commas.

5.1 WHAT DOES A DECISION INVOLVE?

It is important that we clarify what happens when we make a decision in ordinary life as we broach the subject of decision making for a computer. We need *to develop a model* of decision making *that is consistent with the way a computer "makes" a decision*. If we practice developing and applying this model in everyday decisions, then we will be better able to adapt our thinking to write the code that allows the computer to "make" a decision.

DEVELOPING A MODEL FOR MAKING A DECISION

There are many situations in ordinary life where we make a decision, which is making a choice between two or more options. Once the decision is made —an option is chosen—we may have a resulting course of action associated with that choice. Let's review the structure of a decision in everyday life and later we will apply this structure to programmed decisions.

When you make a decision you must choose *one of at least two* things. Once the choice has been made, you may be required to follow a specific course of action for that choice. We will call each resulting course of action an *outcome*.

Consider the decision of whether to buy tickets to a concert. The decision is a *choice* between *two options*: buying the tickets or *not* buying the tickets.

Decision	*Option 1*	*Option 2*
	buying tickets	not buying tickets

Now what do we mean by an outcome? An outcome is a resulting course of action associated with each option. What is the resulting course of action of choosing Option 1? The resulting action could be that you work overtime to make extra money, you get an evening off from work for the night of the concert, you then go to the ticket agency, and you spend the money on the tickets. The resulting action of choosing Option 2 could be nothing. You don't have to do anything if you decide not to buy tickets to the concert.

Outcome 1	*Outcome 2*
work overtime	nothing
get a night off	
go to ticket agency	
buy the tickets	

Consider another decision from everyday life that involves *two options* and *two outcomes*. Your parents are going away for the weekend. They ask you whether you would like to go away with them to the mountains. If you go to the mountains, you can go hiking or sailing on a nearby lake. If you stay home, you'll be all alone, but your mother wants you to paint the porch while they are away. So your decision is to either hang with your parents for the weekend or work like a slave while they are away. Let's consider the decision, the options, and the outcomes associated with this situation.

Decision	*Option 1*	*Option 2*
	go away	stay home
	with parents	alone
	Outcome 1	*Outcome 2*
	ride in the car with parents	wash porch
	eat dinner with parents	sand porch
	go hiking	paint porch
	go sailing	have friends over if you can stay awake

Depending on which option you choose, an outcome will follow, if it exists for that option. The important thing to notice about decisions is what your options are as well as the outcomes for each option. Decisions always involve a choice between at least *two* options. Once you decide, you then branch off to follow the outcome associated with that decision (Figure 5.1 A and B).

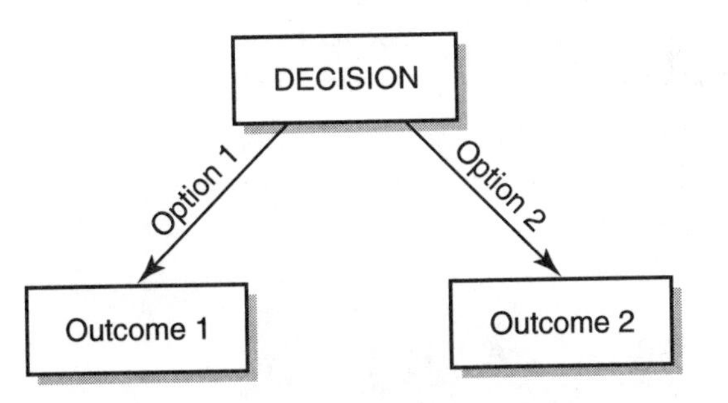

FIGURE 5.1A A decision involves a choice leading to two outcomes: Outcome 1 and Outcome 2. Outcome 1 leads to the left branch and Outcome 2 leads to the right branch.

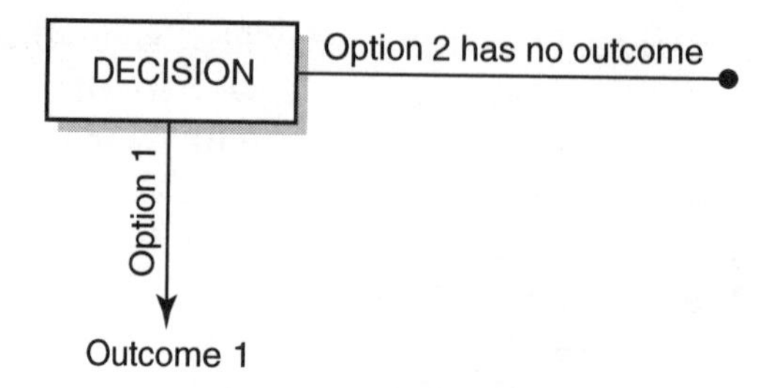

FIGURE 5.1B A decision involves a choice leading to only one outcome. The other choice has no resulting action.

Let's look at some other examples from everyday life. Put each decision in the context of our model using outcomes. Consider what the decision is in each situation—what your options are—and whether each option has an outcome.

EXAMPLE 1

If you go to the early showing of a movie, you will be home in time to watch a program you like on TV. If you choose the later show, then you should set the VCR timer to record the program (Figure 5.2).

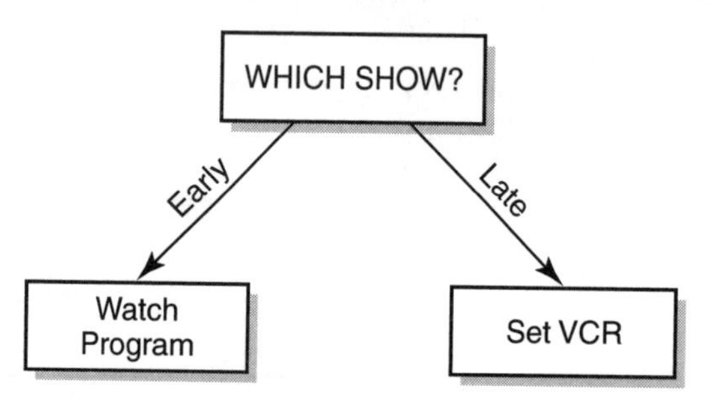

FIGURE 5.2 The choice of the earlier or later show involves two outcomes: watching the program when you return or programming the VCR now.

EXAMPLE 2

Another decision is how you wish to spend your allowance. If you spend it on a shirt you like, you will not have money for a CD. So you must make a decision (Figure 5.3).

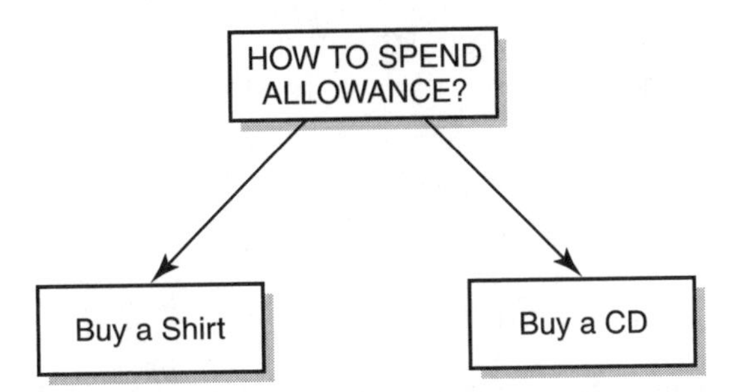

FIGURE 5.3 The decision about how to spend your allowance shows two options—choosing either the shirt or the CD—and two outcomes—purchasing the shirt or purchasing the CD.

A decision, by its nature, always has a choice between at least two things—*this or that, yes or no, true or false*. A decision must involve two options. This is true in our ordinary experience. Computer programs simulate that process.

Remember that a computer is a machine—not a thinking being. It has a limited ability to "decide." This is a machine that only "understands" two things—current "on" or "off," which can be associated with Integer 1 or Integer 0. And now we take this to a higher abstract level where the Integers 1 and 0 are used to represent the boolean values *true/false*, respectively. Don't worry about how this is done—this is done by low-level programmers. The computer will "make" a decision based on whether it gets a boolean value of *true* or *false*. For us, as programmers, we need to develop boolean expressions that model our decisions. So we have to make certain adjustments in our intuitive processes to fit this model of a decision with its options and outcomes *clearly stated*.

As you might imagine, the computer has a limited capacity in how it makes decisions.

When a computer makes a decision, it evaluates a boolean expression.

We use the boolean type because of its values: *true* or *false*. The computer will choose between true and false—and that's it! The computer *always chooses true*. Through the use of a special programming statement, the computer can be manipulated to execute an outcome.

APPLYING THE DECISION MODEL

Let's examine situations where decisions are made by a computer. In each example, we will look at the decision, the options, and the outcomes. As you read each situation, try to identify the decision and outcomes. The decision will be framed so that a boolean expression can be used.

EXAMPLES

Entering a Password at an ATM

Initially, this may not seem to involve a decision, but for the computer it involves a lot of decisions. When you enter a password at an ATM, the computer must match your entered password with the information obtained from the magnetic strip on your bank card. The password *you type* after inserting your card should match the password obtained from the strip. If there is no match then a message will be printed to the effect of "Your password is incorrect. Please enter it again." (see Figure 5.4).

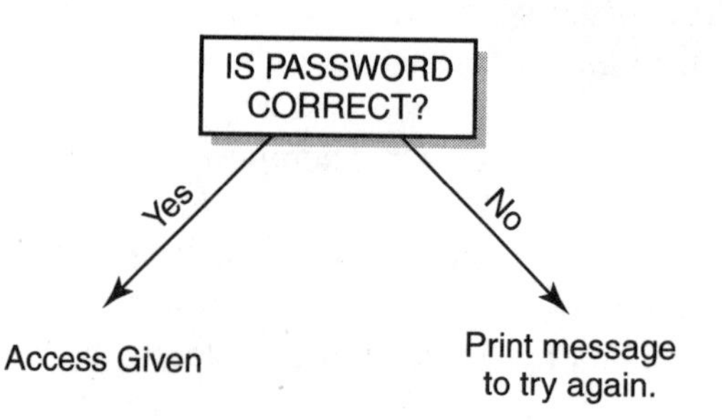

FIGURE 5.4 A password is evaluated for accuracy. If the password is correct, access is granted. Otherwise, a message to try again is given.

Decision: Does the password entered equal (= =) the actual password?
Outcome 1: Grant access.
Outcome 2: Print message saying try again.

Counting the Number of People with Last Names Beginning with "L"

If we want to keep a tally or count of people whose last names begin with the letter 'L,' we need to program the computer to look at the first letter of a name to see if it is the letter 'L.' Every time a name is entered at the computer, we either increase the count of the names or ignore the name because it does not satisfy our condition. So a decision has been made and two outcomes are possible: to count or to ignore (Figure 5.5).

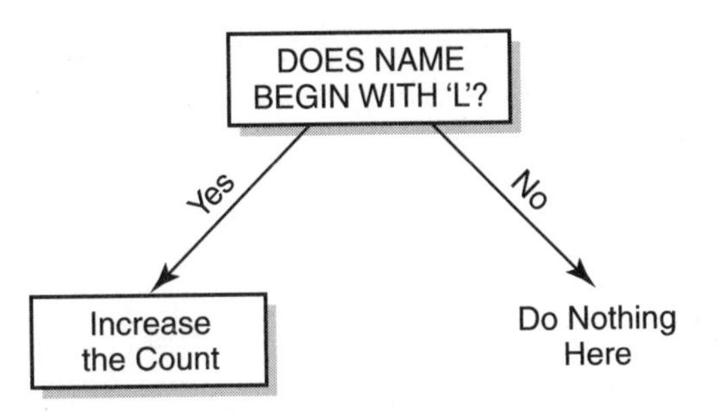

FIGURE 5.5 The name is entered after which a decision is made by the computer to see whether the name begins with an 'L.' If it does, the count of names beginning with the letter 'L' is increased. If not, no action is taken.

Decision: Does the first letter of the name equal (= =) the letter 'L'?
Outcome 1: Increase the number that keeps track of the count of last names beginning with 'L.'
Outcome 2: Nothing.

Heads or Tails

Another example of a decision is the one involved in "heads you win, tails you lose." If you flip a coin and the heads appear, then you have won. The other option is that the tail appears and you have lost (Figure 5.6).

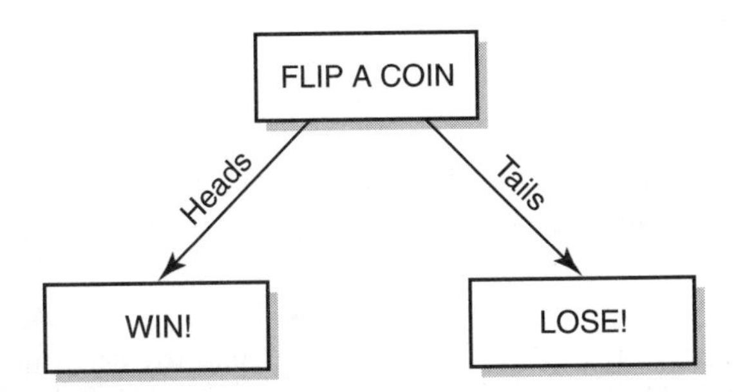

FIGURE 5.6 The outcomes after flipping a coin are shown: winning or losing.

Decision: Does the coin face equal (= =) a head?
Outcome 1: You win.
Outcome 2: You lose.

Decisions always involve a choice between two or more options. Each choice may have an outcome associated with it.

5.3 CONTROLLING WHERE THE COMPILER GOES

It is important to remember that the compiler moves through a program and executes statements in the order in which they are written. If you want to alter that order—and this is important for decision making—you need to understand how to control the compiler.

PROGRAM FLOW

Program flow is an important concept. It is based on the idea that the translator of any program we write in a high-level language will *sequentially* carry out the instructions we write unless programmed otherwise. What does this mean? This means your programming statements are *executed in order—one after the other*.

Look at this example from the programming language B.A.S.I.C. Notice how the program has line numbers to the left of each statement. This numbering of lines emphasizes the *order* of translation of each statement. It helps the programmer to remember that what is on line 10 is executed before what is on line 40.

```
10  LET x = 7
20  LET m =  2 * x
30  PRINT "The answer is"; m
40  END
```

Everything is done *sequentially* unless we direct it otherwise. (The translator here is actually an interpreter, translating one line at a time.) So in the B.A.S.I.C. program, the interpreter goes from line 10 to line 20, and from line 20 it goes to line 30, and, finally, to line 40.

The compiler, another type of translator, is similar to a large ball at the top of a hill. Once you give the ball a little push (or start to "run" a program), it will come down the hill and just roll over everything in its way. The compiler is similar in that it won't stop for anything unless it is *programmed* to do so, because at a lower level it is programmed to *go to the next statement*. It automatically goes there unless the programmer controls the compiler through a *control statement*.

CONTROL STATEMENTS

A *control statement* in a programming language is a statement that *allows the compiler to alter its normal execution of line by line execution of code*. Certain control statements allow the compiler to *skip over one or more lines* of code to get to another line or lines of code. Some control statements allow the compiler to *repeat certain lines* of code. The ability to skip a line or block of code is important when you write a decision because you want to be able to *execute* one outcome from all the outcomes that follow the decision. You don't want to execute *all* the outcomes but, rather, you would like the compiler to *go to the outcome that it should execute and skip all the others*.

Our first control statement will allow the compiler to evaluate a boolean expression and then either go to the next line or skip that line. To understand how a control statement works, we need to look at a specific control statement—*the if...statement*.

5.4 THE IF...STATEMENT

The *if...statement* is our first example of a control statement in a programming language. An *if...statement* has two parts: a *boolean condition* followed by a *conclusion*.

The Boolean Condition

A *conditional* statement (you might have seen one in a Geometry class) is a statement that begins with the word "if." A *boolean condition* is a conditional statement containing a boolean expression. Another name for a conditional statement is a *hypothesis*. In computer programming languages, a hypothesis is formed by using the word "if" with a boolean expression. The boolean expression can be evaluated as *true* or *false*. When the boolean expression within the hypothesis is *true*, then the conclusion occurs. The conclusion will not happen unless this hypothesis is "satisfied" (i.e., the boolean expression is true). So *the if...statement* uses the boolean expression as a way of deciding whether the conclusion that follows is executed. If the boolean expression is *true*, the conclusion is executed.

The Conclusion

The *conclusion,* another name for an *outcome,* is a statement that follows the hypothesis. If the hypothesis is *true* then the conclusion *occurs because it is executed by the computer.* The conclusion represents one outcome you would like to have happen when the hypothesis is *true.* In the examples that follow, each underlined statement is a conclusion.

EXAMPLES

If it rains today, we won't go outside.
If I have enough money, I will order tickets.
If the password is correct, I will get access to my account and withdraw money.

EXAMPLES OF THE IF...STATEMENT IN EVERYDAY CIRCUMSTANCES

Hypothesis/Boolean Condition	**Conclusion**
If it rains tomorrow,	we won't go.
If you win the game,	you advance to the next round.
If they get home by 9,	we can leave by 10.

Let's rephrase each hypothesis with its boolean expression underlined. In order for any of these conclusions to occur, you need to ask whether the boolean expression of the hypothesis is *true.* In a sense, each statement can be rephrased like the following:

Boolean Condition	**Conclusion**
If it's true that it will rain tomorrow,	we won't go.
If it's true that you won the game,	you advance to the next round.
If it's true that they'll get home by 9,	we can leave by 10.

EXAMPLES OF THE IF...STATEMENT FOR A COMPUTER

Here are some examples of *if...statements.* Remember that the boolean expression is contained within the part that begins with "if." The boldface part is the conclusion.

EXAMPLES

if amount of the check is less than the balance
boolean expression
subtract the amount of the check from the balance.
conclusion

if password entered at the keyboard is the same as the true password
boolean expression
provide access to the account.
conclusion

if your age is greater than 16
boolean expression
apply for your driver's license.
conclusion

In each example, the hypothesis can be rewritten using a boolean expression. Boolean expressions, if you recall, come from using the *relational operators*: < (less than), > (greater than), <= (less than or equal to), >= (greater than or equal to), == (equal to), and != (not equal to). Let's take each of these conditions and rewrite it using relational operators to create a boolean expression. In each example, we will declare any variables we need so that we can write a boolean expression.

EXAMPLES USING THE RELATIONAL OPERATORS

I.

double check_amount, balance;

if amount of the check is less than the balance
boolean expression
subtract the amount from the balance.

if *(check_amount < balance)*
boolean expression
subtract the amount from the balance.

II.

string entered_password, real_password;

if data entered at the keyboard is the same as the actual password
boolean expression

provide access to the account.

if (*entered_password == real_password*)
boolean expression
provide access to the account.

III.

char my_char;

if the first letter of the name is an 'L'
boolean expression
increase the number of these names.

if (*my_char* == 'L')
boolean expression
increase the number of these names

IV.

int age;

if your age is greater than 16
boolean expression
apply for your driver's license.

if *age* >16
boolean expression
apply for your driver's license.

The compiler through the *if...statement* "makes a decision" based on the value obtained from the boolean expression. A value of *true* allows the compiler to execute the *conclusion* of the *if...statement*. A value of *false* allows the compiler *to skip the conclusion* to go directly to the next statement. The important fact to remember about the *if...statement* is that the line immediately following the *if...statement* will be executed *no matter what*. The only alteration for the compiler is whether it *skips over the conclusion* after the hypothesis.

What the *if...statement* allows you to do, in terms of the decision model we discussed earlier, is to put *one outcome* as the *conclusion* of the *if...statement* (Figure 5.7).

TIP!

The *if...statement* needs a boolean expression to work properly.

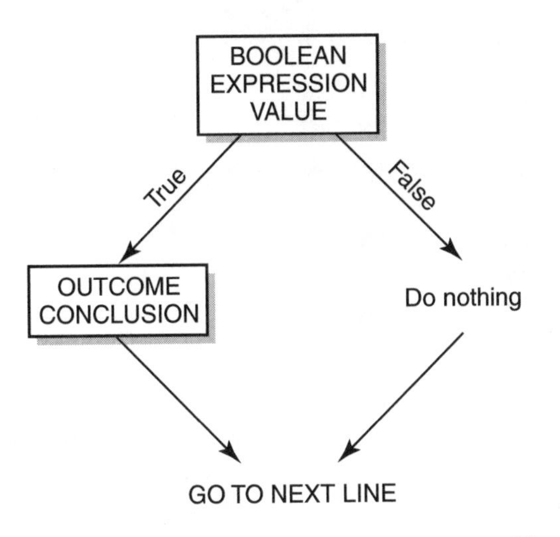

FIGURE 5.7 A model of the *if...statement* with one outcome.

If there is *more than one outcome* from a decision, you need to use another control statement. There is no place for a second outcome to be executed when using the *if...statement*. The only second outcome that "works" in the *if...statement* is to do nothing.

A Block of Code

At this point, we need to introduce how to create a block of code—a group of programming statements that *should be executed as a group*. We use braces to block off a group of programming statements: { }. The first time we saw these braces was when we introduced the main section of a program in Chapter 4. Braces are a way of instructing the computer that we are *sectioning off* some lines of code.

With the control statement we just studied—the *if...statement*—we need to be able to write code for an outcome that has *more than one instruction*. Let's say you wish to ask a person playing a video game whether she wishes to continue playing the game. If she says "yes," you need to start the game again *and* reset the score.

If answer = = "yes"
    Outcome: {start game again
        and set score to zero.}

In another example, you will access your bank account only if your password is correct. If it is correct, you'll be granted access *and* asked what you would like to do next—make a deposit, get cash, and the like.

If password entered at keyboard = = password obtained from magnetic strip
Outcome : {provide access,
ask what user wishes to do,
and get input from her.}

HINT!

Whenever *more than one* statement is used in the context of a control *statement*, use the braces to make a block of code.

5.5 THE IF...ELSE...STATEMENT: THE TWO OUTCOME DECISION

The *if...else... statement* is another example of a *control statement*. Like the *if... statement* it uses a boolean expression followed by two conclusions (two outcomes). These two conclusions are executed according to the *value* of the boolean expression. Let's first look at some examples of the *if...else...statement* in everyday life.

EXAMPLES

If Tom has to work tomorrow, I'll see him tonight; otherwise, I'll see him tomorrow.
If the repairs cost more than $2000, I'll buy a new car; otherwise, I'll get it repaired.

Each example, upon analysis, has a boolean condition and two outcomes. Let's identify the decision and outcomes for each example.

I.

Decision	*Option 1*	*Option 2*
	Tom works tomorrow.	Tom doesn't work tomorrow.
	⇓	⇓
	Outcome 1	*Outcome 2*
	see him tonight	see him tomorrow

II.

Decision	*Option 1*	*Option 2*
	repairs cost > $2000	repairs cost < = $2000
	⇓	⇓
	Outcome 1	*Outcome 2*
	buy a new car	repair the car

Now let's take each example and fit it into the *if...else...statement* structure. Depending on the value of the boolean condition, one of the two outcomes will be executed.

EXAMPLES

If Tom has to work tomorrow, I'll see him tonight *else* I'll see him tomorrow.

boolean expression outcome 1 outcome 2

If the repairs cost more than $2000, I'll buy a new car *else* I'll get it repaired.

boolean expression outcome 1 outcome 2

Boolean Expression	*Value*	*Which outcome is executed?*
Tom works tomorrow.	true	Outcome 1: I'll see him tonight.
Tom works tomorrow.	false	Outcome 2: I'll see him tomorrow.
The repairs cost more than $2000.	true	Outcome 1: I'll buy a new car.
The repairs cost more than $2000.	false	Outcome 2: I'll get it repaired.

The value of the boolean expression *determines* whether Outcome 1 or Outcome 2 is executed. Whenever the boolean expression is true, Outcome 1 (always placed first) is executed. Whenever the Boolean expression is false, Outcome 2, placed after the word "else," is executed.

EXAMPLE

If the hypothesis/boolean expression is *true*
    execute
    Outcome 1
else

execute
Outcome 2

If the outcomes have multiple programming statements, then you need to use the braces around each *block* of instructions.

EXAMPLE

If (expression)
boolean expression is true

execute

Outcome one's {first statement,
second statement,
third statement, etc.}

else

execute
Outcome two's {first statement,
second statement,
third statement, etc.}

SOME EXAMPLES IN CODE

To save time, we will often examine only portions of a program—these are called *program fragments*. Rather than show the compiler directives and the main section for every example, we will just show *a portion of the program* that we need to study.

Here is a fragment from the programming language C++; an example of the *if...statement* is written this way:

```
if (x > 12  )
  cout << "The variable is greater than 12." << endl;
```

Note that *parentheses are used to surround the boolean expression* in this language. Consider these examples.

EXAMPLES

If a number > 0 then inform the user that the number is positive
else

inform the user that the number is *not* positive.

```
if (number > 0  )
     cout << "The number is positive." << endl;
else
       cout << "The number is not positive." << endl;
```

EXAMPLE

If your age >= 16 then inform the user she is old enough to drive
else
inform the user that she is not old enough to drive.

```
if (age >= 16   )
     cout << "You are old enough to drive." << endl;
else
     cout << "You are not old enough to drive." << endl;
```

USING A BOOLEAN CONDITION TO DETERMINE WHETHER A NUMBER IS "EVEN" OR "ODD"

In Examples one and two, we will let the computer evaluate whether a number is "even" or "odd." To do this, we need to use the *mod* operator (gives a remainder in division) from Chapter 3. Let's first review a few examples with the *mod* operator (%).

18 % 2 produces 0 because 18 divided by 2 is 9 with no (0) remainder.
15 % 2 produces 1 because 15 divided by 2 is 7 with a remainder of 1.

18 is an even number. *All even numbers will have no remainder when they are divided by* 2. So we write a boolean expression using the equality relational operator, = =. If a number is even, dividing the number by 2 gives us 0. If a number is odd, dividing the number by 2 gives us 1. Consider these examples using the variable x that has been assigned a value.

```
int  x;
x = 24;

x % 2 == 0     // is 24 % 2 == 0?
   ⇓
```

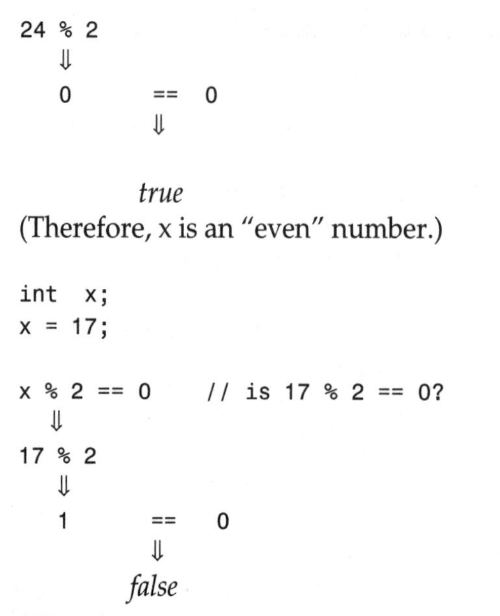

```
24 % 2
   ⇓
   0        ==   0
            ⇓
```

true

(Therefore, x is an "even" number.)

```
int  x;
x = 17;

x % 2 == 0     // is 17 % 2 == 0?
  ⇓
17 % 2
   ⇓
   1        ==    0
            ⇓
```

false

(Therefore, x is not "even" and must be "odd.")

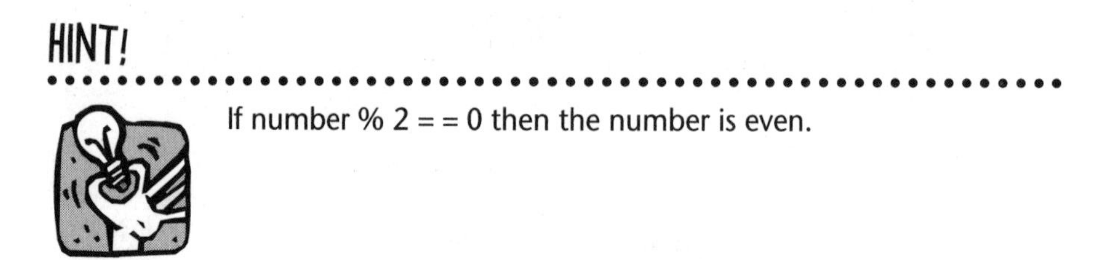

HINT!

If number % 2 = = 0 then the number is even.

HINT!

If number % 2 = = 1 then the number is odd.

Now we are ready to use the mod operator (%) in the context of an *if...statement*. Both fragments accomplish the same thing—one uses a boolean variable (a holder for the boolean value) and the other does not.

EXAMPLE 1

If a number is even, print a message saying it is even.

Program fragment with a boolean value:

```
int number;
cin >> number; // let the user give us the number
if ( number % 2 == 0)
  cout << "The number is even."<< endl;
```

Program fragment using a boolean type variable:

```
int number; boolean  answer; //both variables are declared

cin >> number; // let the user give us the number
answer = number % 2== 0;// answer holds true or false
if (answer)
  cout << "The number is even."<< endl;
```

Notice that the last statement uses a boolean variable in place of a boolean expression. This is possible because the value inside the variable, *answer,* will determine whether the conclusion is executed.

EXAMPLE 2

If a number is even, print a message saying it is even; otherwise, print a message saying it is odd.

Program fragment with a boolean value:

```
int number;
cin >> number; // let the user give us the number.
if ( number % 2 == 0)
  cout << "The number is even."<< endl;
else
 cout << "The number is odd."<< endl;
```

Program fragment using a boolean type variable:

```
int number; boolean  answer; //both variables are declared

cin >> number; // let the user give us the number
answer = number % 2 == 0;//answer holds true or false
if ( answer)
  cout << "The number is even."<< endl;
else
 cout << "The number is odd."<< endl;
```

5.6 THE SWITCH/CASE STATEMENT

Decisions are generally made between two options. If you want to decide between *more than two* options, you can use a "switch" statement, as it is called in the C++ programming language. Other languages use a different name. Once you understand what this statement does, you will be able to recognize it no matter what it is called in any language.

A "switch" statement works in this way. Think of this analogy: You are standing in a huge room that has three doors in it. Each door has a number on it. If you unlock Door # 1, Outcome 1 is behind the door. If you unlock Door #2, Outcome 2 is behind the door. The same thing is true for Door #3. Behind each door is a different set of instructions that comprises that outcome.

The "switch" statement evaluates an *integer* variable—that is, looks at its *value*. Then it examines a list looking for that value. When it finds the value, it executes all the instructions that are associated with that value. This type of control statement always has *two* elements. One element is where an integer variable is examined to see what value is inside it. Then the value is found among a list of integer values and the statements (the outcome) next to that value are executed. Here is a diagram using our "door" analogy:

switch Door_Number
Door #1: Statement 1; Statement 2; Statement 3;
Door #2: Statement 1;
Door #3: Statement 1; Statement 2;

The actual "switch" statement would need to "know" that the variable used to control it is an integer, so we first declare an integer. Also, it uses the word "case of " instead of Door # to list the integers. Let's do an example from a program that could be used at an ATM machine. The user will be asked whether she wishes to make a deposit, get cash, or check the balance of an account. If she inputs a "1" then she will expect to be able to make a deposit. If she inputs a "2," she will be able to get cash, and if she inputs a "3," she will be able to examine the balance of one of her accounts. In the program fragment that follows, the numbered statements represent *undisclosed* programming statements.

```
int your_choice;

cout << "Please choose your option by typing the number 1, 2, or
3." << endl;
```

```
cin >> your_choice;


switch  (your_choice)
{
case 1:  Statement 1; Statement 2; Statement 3;

case 2:  Statement 1;

case 3:  Statement 1; Statement 2;
}
```

Not all outcomes for an integer have the same number of statements. One outcome might have only one statement. Another outcome might have three statements. The "switch" statement is a clean way of choosing among *more than two things*.

As you learn more about decisions, you will see that they can be more involved than the models we examined here. There are interesting ways to handle more complicated decisions. It just takes some practice to use the statements we have already mentioned in the correct way. The *if...statement* with the *if...else...statement* and some variations on those two statements—including using them together or one or both of them repeatedly will allow any programmer to handle complicated decisions having more than two choices and two outcomes.

SUMMARY

The boolean type is another type of variable in addition to those we have already studied—the integer, character, string, and real. It is used to store the result of evaluating a boolean expression. It always holds the value *true* or the value *false*.

We examined how a decision can be modeled on options and outcomes. If you choose one option, it leads to one outcome. If you choose another option, it leads to another outcome.

Control statements are programming statements that allow the compiler to alter the usual order of execution of statements. In our case, we want to skip over one line of code to get to another. One example of a control statement, the *if...statement*, has two parts—a hypothesis or boolean condition followed by a conclusion. It is used, ideally, for a decision that has only one outcome.

The *if ...else...statement* has three parts—a hypothesis containing a boolean expression followed by two outcomes—only one of which will be executed depending on the value of the boolean condition. If the boolean expression is *true,* the first outcome is executed. If the boolean expression is *false,* the second outcome is executed.

The *if ...statement* can be used to determine whether a number is "even" or "odd." This useful algorithm is done through the mod division operator (%). Finally, a *switch* statement is a statement that is best applied to decisions that involve more than 2 choices.

ANSWERS TO ODD-NUMBERED EXERCISES

Ex. 5.3

1. false 3. false 5. true 7. false

CHAPTER

6

Loops or How to Spin Effectively

IN THIS CHAPTER

- We Introduce the Loop
- The Counter Statement
- Different Types of Loops
- Fixed Iterative vs. Conditional Loop
- The For Loop
- The While Loop
- The Do...While Loop
- Examples with Loops

Many times in the course of programming, you wish to use the computer to repeatedly do some task. This involves constructing a *loop*. In this chapter, we examine how a loop encloses a group of lines and repeats them a certain number of times. There are different kinds of loops and we will explore each type.

6.1 WHAT IS A LOOP?

Visualize a cowboy using a rope to lasso a dog's neck, a circus ring, or the balloon that encloses a cartoon character's thoughts. Each of these is a loop. A loop *always encloses* something. When you think of a loop, some of these images might come to mind (Figure 6.1).

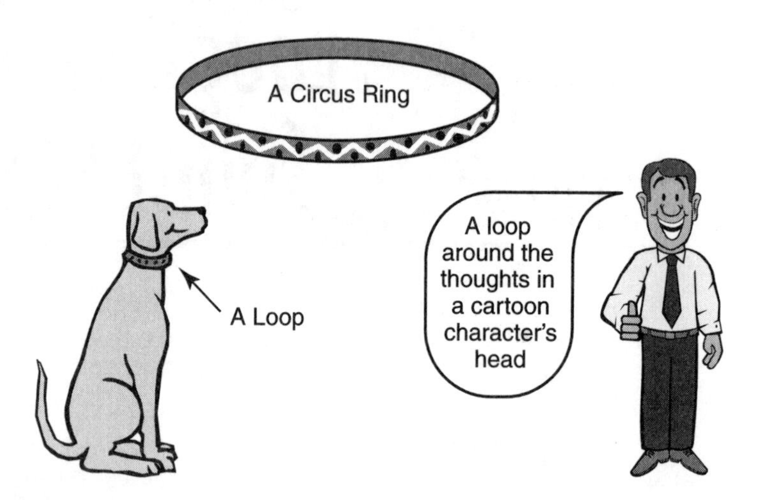

FIGURE 6.1 A loop around a dog's neck, a circus ring, and a loop around the thoughts in a cartoon character's head are all shown.

A *loop* brings to mind a circular image. What does the term "loop" have to do with computer programming? We use the term to describe *a part of a program that you execute over and over again* until you are permitted to leave that part to get to another programming statement. In a sense you continuously *loop back through statements that you just executed* and you execute them again.

Another image of the *loop* is one of going back to where you once were. Let's say you leave your house and go to your friend's house, but your friend is on her way to your house. You'll loop back to pick up your friend. In this sense, a loop means *to go back to a place* where you once were. "To loop back to something" means to go back to where you were and start again (Figure 6.2).

FIGURE 6.2 A person drives a car back home ("loops back") to pick up a friend who is standing next to the house.

LOOPS IN PROGRAMS

In a program, a loop describes a group of one or more lines of code that must be repeated some number of times. Consider this example: We are going to write a program to get 10 numbers from the user and check whether each number is "even" or "odd." We need to get one number at a time from the user before checking to see whether it is "even" or "odd." Then we go back to *get another number to do the same thing*. As we repeat each step, we *go back to where we were* in the program to repeat the *same* instructions. This is a loop. This loop encompasses two steps: getting the number and evaluating it as either "even" or "odd" (Figure 6.3).

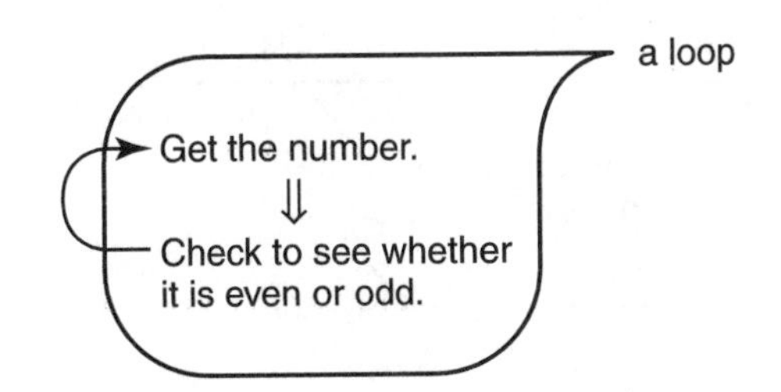

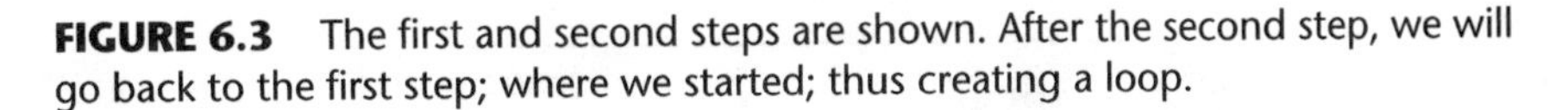

FIGURE 6.3 The first and second steps are shown. After the second step, we will go back to the first step; where we started; thus creating a loop.

Another way to envision a loop is to see it as a *circle* around a few lines of code. That way we have a clear understanding that everything within the loop (circle) will repeat as long as it should. We still have not mentioned anything about controlling a loop—for example, when does it end? How long does it last? These are natural questions that you should have when you first start learning about loops and they will be answered shortly.

ANOTHER CONTROL STATEMENT

The loop is another example of a *control statement*. When you stay in one block of code (a group of programming statements) to do something over and over again, you are *controlling program flow*. Just as we mentioned in Chapter 5, normal program flow is *in sequence*—executing one line after another. A loop forces program flow to stay on a certain line or in a block until allowed to exit that place and proceed *in sequence*. We will soon discuss how a loop is *entered* or *exited*.

REPEATING STEPS IN A LOOP

We make a decision to use a loop when we realize we want to do something over and over again on the computer. If you want to write the sentence, "I am sorry for chewing gum during class" one thousand times, your best decision would be to write a program that will print that phrase repeatedly. If you want to add 20 numbers, then you want to use a loop to get one number at a time and then add that number to the tally of all the numbers so far (Figure 6.4).

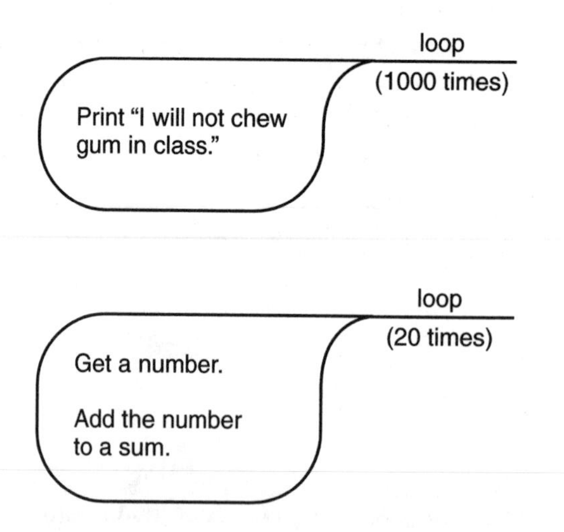

FIGURE 6.4 Each programming statement is shown enclosed by a loop.

There are many examples of situations that use loops. Every time you enter an incorrect password at an ATM, the machine asks you to enter it again. The machine is actually *executing a loop* which, upon getting incorrect data, prints the message "Your password is incorrect. Please type it again." It will *repeatedly print this message* until you give it the correct password or it swallows your card. You can only get out of the loop by either typing the correct data or by the machine putting a limit on your number of attempts (Figure 6.5).

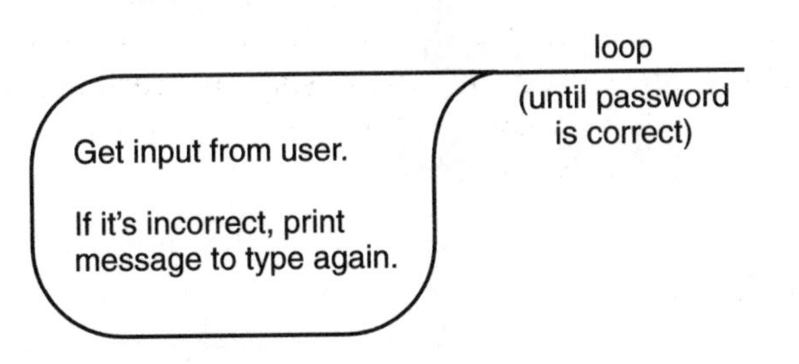

FIGURE 6.5 A loop is shown surrounding the body of instructions in the ATM situation.

A COUNTER STATEMENT

To understand how loops work, we need to learn a preliminary statement that is used in most loops, which is called *a counter statement*. A *counter statement* is a statement that allows a variable to increase its own value— usually by 1. For example, if a variable *x* has the integer 3 inside of it, after a counter statement is used with *x*, *x* will contain the value 4.

How is this done? We write a statement to take the variable (in this case, *x*) and add one to its present value. Then the statement assigns the new value to the variable *x*. So the statement has two parts—one that increases the value of *x* by 1 and another part that assigns that value to *x*. This is what it looks like:

x	=	x + 1;
left-side	operator	right side

Think of the *right side* of the statement first. Adding 1 to *x* just obtains the next integer greater than *x*. Assigning it to *x gives* x *that new value*. One is added to whatever is inside of *x*. If 3 is inside of *x*, then 3 + 1 is 4. Now the value 4 is really what is meant by "x + 1" so 4 is *assigned to x*.

x = x + 1;
(3 + 1)
⇓

x ⇐ 4

Another way to understand this statement is to consider the precedence of operators (Chapter 2). The arithmetic operators (+, –, *, /) have *higher* precedence than the assignment operator (=). So, 1 is added to *x* first, and then the result is assigned to *x*.

THE USE OF THE COUNTER STATEMENT

A counter statement is used frequently in the context of loops to count the number of times a loop spins. Let's see how this is done. This statement is used in the same way that a ticket taker works at the gate of a concert. Every time a person walks through the gate, the ticket taker clicks the clicker to keep a count of the number of people who pass through the gate.

Count = Count + 1;
"new" Count = "old" Count + 1

Another way to think of the counter statement is to think of the variable on the left side as being the "newest" Count. The "newest" Count we have is the most recent value of the variable. It should get what is inside of Count (now) with one more added to it. "New Count gets old Count plus one" (Figure 6.6).

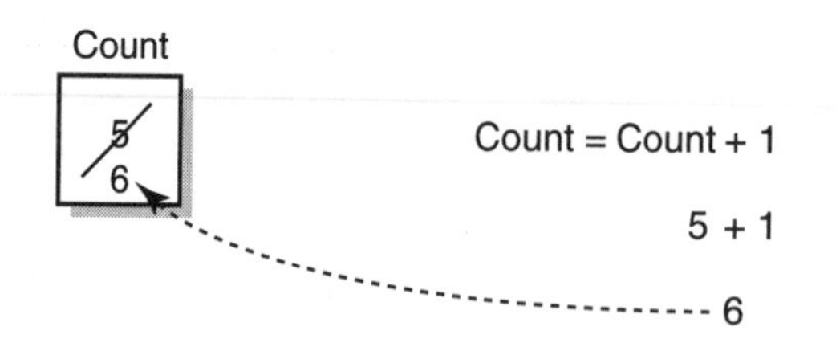

FIGURE 6.6 Count is originally shown with the value 5. The addition (+) is executed before the assignment (=) so that 6 gets assigned to Count, wiping out the old value of 5.

6.2 TWO DIFFERENT KINDS OF LOOPS

There are two different kinds of loops in most languages. The main difference between loops is whether the loop is programmed to "spin" (a run through a loop) a *known* or *fixed* number of times. Some loops can be designed to spin 15, 30, or 100 times—for example. The programmer knows *exactly how many times* he wants to execute the loop. Other loops spin an unknown number of times *at the time of writing the loop*. These loops spin until something "happens" to stop them.

FIXED ITERATIVE LOOPS VS. CONDITIONAL LOOPS

If you want to print a message 100 times on the screen, you use a *fixed iterative loop*. The term "fixed" implies that the number of times the loop spins (100 times) is known at the time the loop is written. The word "iterative" means "repeating." If you want to send a file to 50 people on a list, you know you will be sending the file out 50 times—that task should use a fixed iterative loop that will execute 50 times.

An example of a loop that spins *an unknown number of times* would be a loop that allows a player to play a computer game over and over again until he indicates he wants to stop playing. The programmer can't know ahead of time when the user will get sick of the game and want to stop playing. The player is always asked at the end of a game, "Do you want to continue?" Depending on the player's answer, the game playing ends (i.e., the loop stops spinning). This is an example of a *conditional* loop. The loop spins until something "happens"— or, as we shall see shortly, until a boolean condition becomes *false*.

In another example, a loop controlling the number of times a game is played could depend on *whether the player wins or loses*. The loop stops when the player loses. As the programmer, you won't know ahead of time when the player will lose a game and be forced to stop.

In both cases the loop was exited when an event *happened*—that is, the player didn't want to play anymore or he lost. The loop spun until something "occurred" to stop its spinning. We either know ahead of time how many times we want to do something or we don't. Depending on our type of situation, we choose the appropriate type of loop.

The first loop to examine is the *for loop*. The *for loop* should be used when you *know how many times* you want to do something.

6.3 THE FOR LOOP

The *for loop* is a *fixed iterative* loop. Although its syntax varies from language to language, certain aspects of the *for loop* are the same. For one thing, it always has the word "for" mentioned at the beginning of the loop. The other characteristic of this loop is that you can figure out how many times it will spin before it is executed.

The *for loop* has *three key instructions* in addition to the group of statements you wish to execute repeatedly—called the *body* of the loop. The first instruction is to *declare* a variable and *assign* it an *initial value*. The second instruction is to make sure that the variable is within range. (We will compare it with some limit.) The third instruction is to change the value of the variable. Before the third instruction occurs, the *for loop* executes the *body* of the loop.

As an example, let's use a loop to print a message 5 times on the computer. First, take a variable, *y*, and set its initial value to 1. Now compare 1 with the upper limit you have in mind (5). Go into the loop and execute the body of the loop—printing the message. Next take *y* and increase its value from *1* to 2. Now you're on your second "spin." Go back to the second instruction, which is to *compare* the variable with our upper limit, 5. Because 2 is less than 5, we continue and go into the loop to print the message again. Next, increase *y's* value from 2 to *3*. Compare 3 with 5 and continue because 3 is less than 5. (You are only on your third spin of the loop!) This process continues one more time as *y* increases and becomes 4. Just before exiting, the message is printed again and *y* becomes *5*. Then the number *5* (the value in *y*) is compared with *5* (the upper limit). This comparison triggers an exit from the loop because *5* equals *5*, but is not less than *5*. The loop is exited (Figure 6.7).

The *for loop* has the following elements: *a control variable*—it facilitates the entire "mechanics" or "workings" of the loop, *a boolean expression* that checks to see whether the control variable is within range, and *a counter statement* to change the value of the control variable. We will examine each of these elements in order to understand the structure of the *for loop*.

THE CONTROL VARIABLE

A control variable is a variable that does the "work" of the loop. (In our previous example, it was the variable, *y*.) It must be declared first and then given an *initial value* before the loop starts to spin. The control variable is then used in a boolean condition to determine whether the loop

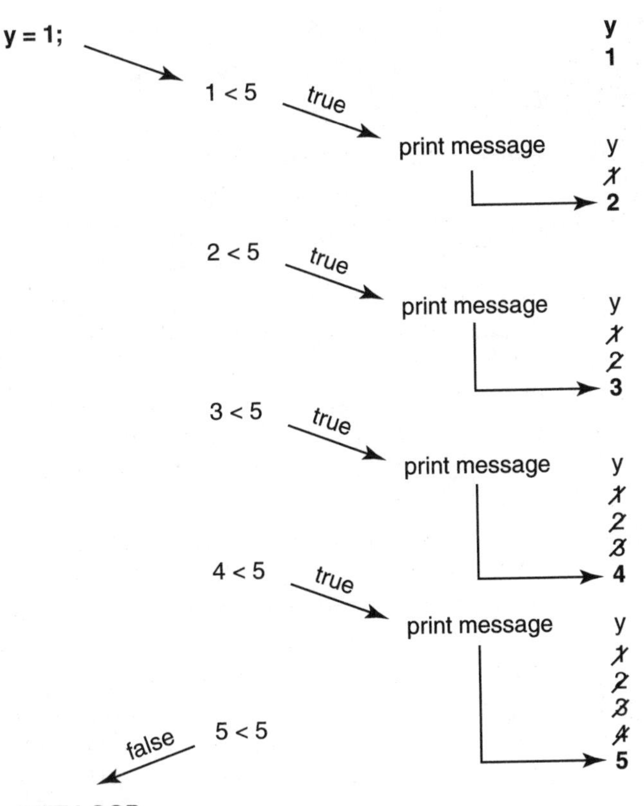

FIGURE 6.7 The variable y is shown with its "old" values repeatedly crossed out and replaced by new values. The comparison statements are shown each time using new values in the boolean expression.

"spins." Just like other control statements, the *if...* and the *if...else...*, the boolean expression must be *true* for anything to happen. The loop will be exited when the boolean expression becomes *false*.

The general idea is that a control variable is *compared with some limit*—an upper limit or a lower limit. When it is *compared* with an *upper limit*, the control variable is checked to make sure that it is still *less than* the *upper limit.*

control variable < upper limit

As long as this is still *true*, the body of the loop is executed. After the body is executed, the control variable is *increased*.

Conversely, the control variable could be *compared* to a *lower limit* in the following manner. The control variable would be checked to see that it is still *greater than* a *lower limit*.

control variable > lower limit

As long as that is *true*, the body of the loop is executed. Then the control variable is *decreased* (Figure 6.8A and B).

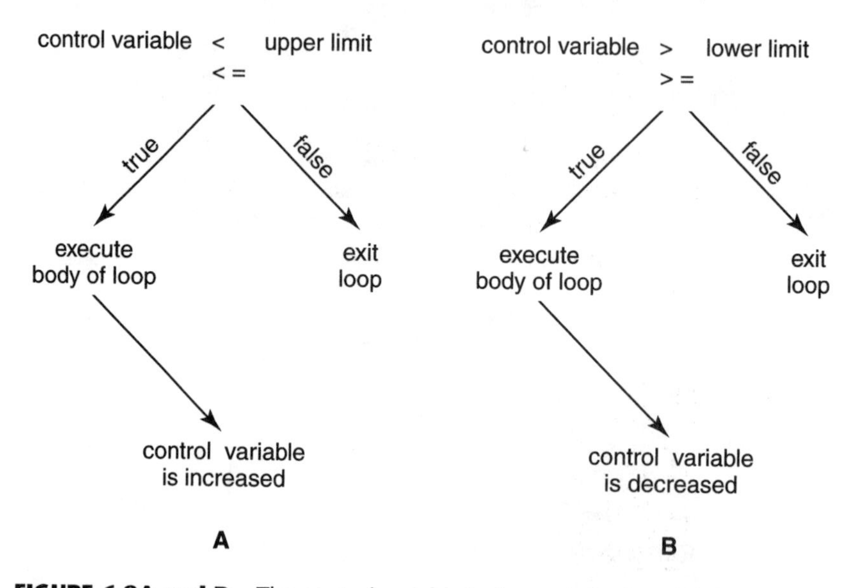

FIGURE 6.8A and B The control variable is shown in each example—compared to an upper limit or a lower limit, respectively. Depending on the value of the boolean expression, either the body is executed and the variable changed or the loop is exited.

Part of the structure of the syntax of a *for loop* is assigning the control variable its initial value—that is, the value it should have at the *start* of the execution of the loop. Then during the loop's repeated executions, the control variable *changes* value. Finally, when it reaches the last value and the boolean expression becomes *false*, the loop is exited.

THE INITIAL STATEMENT

The first statement contained within the parentheses of the *for loop* is the initialization statement for the loop's controlling variable. You must *declare* the variable before giving it its *initial* value in separate statements or in one combined statement. Notice that we have left the rest of the *for loop's* statement blank for now. This syntax (grammar) shows how to assign an initial value to the control variable.

Declared and then Assigned	Declared and Assigned in the Same Statement
`int x;` `for (x = 1;  ... )`	`for  (int x = 1;  ...)`
`int y;` `for (y = 100; ... )`	`for ( int y = 100; ...)`

Notice that in the first example, the initial value of the variable was *1*. It is easiest to start counting at 0 or 1 and this is the reason why most *for loop's* will initialize the control variable to one of these integers. The second example uses a large number as its initial value—you will soon see why.

HINT!

The *for loop*'s control variable is always an integer.

THE BOOLEAN EXPRESSION

The next statement used in a *for loop* is the *boolean expression,* which must be *true* in order for the loop to execute. Think of the boolean expression as *the permission granted or denied* for the body of the loop to be executed. If the boolean expression is *true, then we enter the loop* and execute all the statements within its control. If the *expression is false,* then *we exit* the loop. To leave the loop means to *go to the first line of code immediately following the for loop block.*

```
for ( int count = 0; count < 25; count = count + 1)
{

// body of loop is here

}
// -> first line immediately following is here
```

Let's examine each of the previous examples and write a boolean expression using the control variable. In our first example, we initialized *x* to be 1 and now we will set the expression to show that *x* should be less than 5. As long as *x* is less than 5 we want to execute the loop. It will look like this.

```
for (int x = 1; x < 5;  ...)
```

Here's another example. We will initialize the variable *y* to 100 and use a boolean expression that shows *y* greater than or equal to 0. Notice the use of the "=" sign in the condition. This means that when *y* reaches 0, it will still *spin one more time* because 0 equals 0. In that case, it will execute the body of the *for loop* one last time.

```
for (int y = 100; y >= 0;  ...)
```

HINT

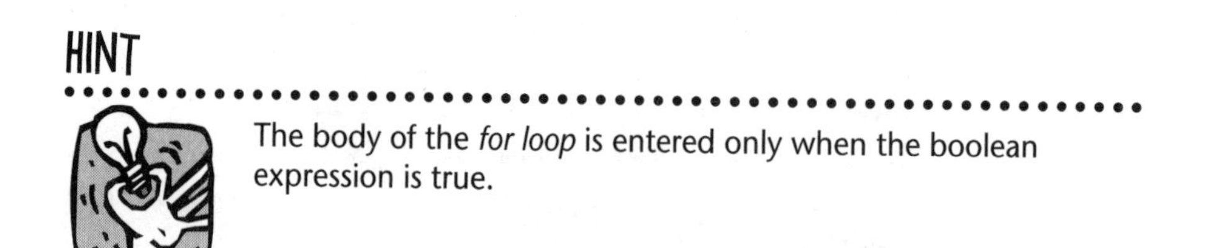

The body of the *for loop* is entered only when the boolean expression is true.

THE COUNTER STATEMENT

The counter statement we mentioned previously is the third statement used in the *for loop*. There are two types of counter statements: one that *increases* the value of a variable and one that *decreases* the value of a variable.

Increases the Variable	*Decreases the Variable*
`x = x + 1;`	`y = y - 1;`
`count = count + 1;`	`answer = answer - 1;`
`age = age + 1;`	`sum = sum - 1;`

Using our previous examples, we insert the third statement, the counter statement, into the *for loop* syntax. The third statement will be executed *after the body* of the loop is executed. Imagine that there are four parts to the entire *for loop*—these three statements and the *body* of the loop. Let's number them in the order in which they happen.

```
for (x = 0;
     1st
                ⇓
           x <= 100 ;
              2nd
     ⇓
{  // body of the loop
     3rd
}
                ⇓
           x = x + 1 )
              4th
```

So the *for loop* executes the first two steps, the 1st and 2nd, *then enters the loop as its 3rd step* and executes the *body*. Lastly, it executes the 4th statement, the counter statement. Keep in mind, too, that the initial statement that sets the control variable's first value *happens only once*. The second time you "loop back" for instructions, the computer will only execute the 2nd, 3rd, and 4th steps (Figure 6.9).

```
    x = 0;        // happens once
      ⇓
  x < = 100;
      ⇓
 body of loop
      ⇓
  x = x + 1;
      ⇓
  x < = 100
      ⇓
 body of loop
      ⇓
  x = x + 1;
      ⇓
  x < = 100
      ⇓
 body of loop
      ⇓
     etc.
```

FIGURE 6.9 The 2nd, 3rd, and 4th steps are shown in sequence repeating each time the loop spins.

Here is another variation on the syntax of a *for loop*. The declaration of the control variable is outside the *for loop*.

```
int y;
for (y = 0; y <= 100; y = y + 1)
```

EXAMPLE

Let's start with an example of a problem that would utilize a *for loop*. These problems are usually ones where *you know ahead of time* how many times you wish to do something. In our example, let's print the numbers 1 through 10 on the screen. The first thing to notice is that we know we want to do something *10 times*—printing those numbers on the screen.

```
for (x = 1; x <= 10 ; x = x+1)
{
cout << x << endl;
}
```

Now let's step through the *for loop* one line at a time—like a compiler would do. First x gets the value *1*, then that value is checked to see whether $x < = 10$ is *true*. Because this is *true,* the body of the loop—only one statement, the printing statement—is executed. The next statement ($x = x + 1$) changes x's value from *1* to 2. After that—recall that the initial statement is not executed again—we go directly to the boolean expression to see that $x <= 10$. (Yes, because $2 <= 10$.) It is true, so we execute the body of the loop and print the number *2* on the screen. After that, x changes from 2 to 3. This pattern continues (Figure 6.10).

EXAMPLE

In a second example utilizing a *for loop,* let's look at a list of 20 numbers and for each number we will print a message on the screen indicating whether that number is "even" or "odd." Here, we will use the *if...statement* from Chapter 5 to help the compiler "decide" whether the number is "even" or "odd."

```
int x;

for ( x = 0; x < 20 ; x = x + 1)
```

Control Variable	Boolean Expression	Value	Resulting Action
x			
~~1~~	1 < = 10	true	1 is printed
~~2~~	2 < = 10	true	2 is printed
~~3~~	3 < = 10	true	3 is printed
~~4~~	4 < = 10	true	4 is printed
~~5~~	5 < = 10	true	5 is printed
~~6~~	6 < = 10	true	6 is printed
~~7~~	7 < = 10	true	7 is printed
~~8~~	8 < = 10	true	8 is printed
~~9~~	9 < = 10	true	9 is printed
~~10~~	10 < = 10	true	10 is printed
11	11 < = 10	FALSE	EXIT LOOP

FIGURE 6.10 The sequence of events in the for loop are shown as the control variable's values are repeatedly changing.

```
{
if( x % 2 == 0 )
 cout << x << " is even. "<< endl;
else
 cout << x << " is odd. " << endl;
/* We were able to use the if ... else... statement from chapter
5. See how useful it is.*/
}
// Did you notice the syntax for the extended comment above?
```

Here is the output from running the above fragment. Each number will be printed with the appropriate tag, "is even" or "is odd."

```
 0  is even.
 1  is odd.
 2  is even.
 3  is odd.
 4  is even.
 5  is odd.
 6  is even.
 7  is odd.
 8  is even.
 9  is odd.
10  is even.
11  is odd.
12  is even.
13  is odd.
14  is even.
```

```
15  is odd.
16  is even.
17  is odd.
18  is even.
19  is odd.
```

You will *not* see "20" on the screen because the loop is exited before 20 is used in the body of the loop. Recall that the boolean condition was "$x < 20$." Once x "hits" 20 or becomes 20, the boolean expression becomes *false,* and the loop is exited.

ON THE CD-ROM!

The complete program that prints even or odd next to a list of numbers.

EXERCISES

Choose the correct boolean expression so that each *for loop* will spin the appropriate number of times.

1. 12 times for (y = 0;   ; y = y + 1)

a) y < 12   **b)** y = = 12   **c)** y < 13

2. 10 times for (m = 1;   ; m = m + 1)

a) m < 10   **b)** m = = 10   **c)** m < 11

3. 20 times for (x = 5;   ; x = x + 1)

a) x < 25   **b)** x = = 25   **c)** x < 26

Complete the third statement for each *for loop* so that the control variable will increase or decrease accordingly.

4. for (y = 0; y < 10;   )

5. for (g = 25; g > 1;   )

6. for (x = 10; x < 100;   )

6.4 THE CONDITIONAL LOOP

A conditional loop is a loop that does not spin a "fixed" number of times. It spins only on the condition that some boolean expression is *true*. You do not know beforehand how many times it will spin. While the boolean expression is *true,* the loop spins. Once the boolean expression changes from *true* to *false,* the loop, *upon finishing the body,* stops.

THE WHILE LOOP

To understand the *while loop*, we first need to look at an analogy. Imagine that you enter a large office room where you must get all this work done. You need to get into the room and clean each desk in the room. You are a very methodical worker so you only clean one desk at a time. This means that you wipe off the desk, sweep under it, and empty the barrel *for each desk* rather than empty *all* the barrels *at once* or sweep under all the desks first.

Another stipulation about how you work is that you can only work in the room *if the light is on*. (It's on an automatic timer.) You go to the door and check to see if the light is on. If it is, you enter the room and start to work. You will work in the room *as long as the light is on*. At some point, you expect the lights to go off because they are on an automatic timer that someone else set. When the lights go off, you *finish the work at the desk you are at and then leave the room.* In a sense, you work while the lights are on. You do each task as outlined in the following description.

while (lights are on: check the lights here!)
{// Enter body of loop to do your work.
Clean the top of one desk.
Pull out the chair.
Sweep under the desk.
Push in the chair.
Empty the barrel next to the desk.}

You have to finish all the steps at one desk even if the lights go off while you're still doing one of the tasks—like sweeping under the desk. Once the lights go out, and you have finished all the work at one desk, you leave the room. However, if the lights are still on, you go on to the next desk. (They actually might have gone off while you were sweeping under the desk. Remember—because of your odd work habits—you can't leave the room until you finish all the work at one desk.)

Now back to our *while loop*. The *while loop* will spin *as long as some boolean expression is true* (analogous to our lights being on). Inside the body of the *while loop*, are the statements you wish to execute. Within the body of the loop there must also be a programming statement that triggers the boolean expression *to change* (like the lights going off with an automatic timer). When that happens, and control passes to the top of the loop, the boolean expression will be false and we will leave the loop. Notice the sequence of arrows indicating how the loop is controlled. Once the last statement in the body of the loop is executed, we "bounce" back up to the top of the loop to see whether the boolean expression is true.

```
    ⇒  while (boolean expression is true)
       ⇓      {
⇑             // the body of a loop
       ⇓      Statement 1;
⇑             Statement 2; etc.
       ⇓      . . .
⇑
    ⇐         }
```

To expand on this loop structure, we need to add some key statements. A counter statement in the body of the loop will allow both loops to function like *for loops*. That is, the number of times they spin can be counted *as they are spinning*. Let's use the analogy to draw a picture of the *while loop's* structure (Figure 6.11).

Let's continue to refine the analogy. When you are in the office you only work while the lights are on. However, you can only check the auto-

while (lights are on)	while (boolean expression is true)
{	{
Work at one desk.	Execute the steps.
⋮	⋮
Did the lights go out?	Did the boolean expression change?
⋮	⋮
Finish work at one desk.	Finish all steps within { }.
}	}

FIGURE 6.11 The *while loop* is shown both in the analogy of the office worker and in a programming context.

matic timer after you finish the work at one desk. The *while loop* operates the same way—you can only check the boolean expression *at the top of* the loop.

This analogy is used to emphasize how the *while loop* does its work. It only checks the boolean expression at the *top* of the loop. For this reason it is called *a pre-test loop.*

A *pre-test loop* is where the boolean expression is checked *first* before entering the body of the loop to execute its instructions. At some point inside the loop, some statement will cause a future evaluation of the boolean expression to be *false*. After the body of the loop is executed, the control goes back up to the boolean expression, and because it is now *false,* the compiler exits the loop and *goes to the next line after the loop.*

THE DO...WHILE LOOP

With the *do...while loop,* the same analogy can be applied. In the office room analogy for the *do...while loop,* you walk inside the room and start cleaning right away *whether the lights are on or off. After you've done the work* (all the tasks for one desk), *you look up to see if the lights are on* in the room. *If they are, you stay in the room* (in the loop) and do the same work on the next desk.

```
do

{
// Enter room to do your work.
Clean the top of one desk.
Pull out the chair.
Sweep under the desk.
Push in the chair.
Empty the barrel next to the desk.}
while (lights are on);
```

The analogy is a little different from the one used in the *while loop* in that you only look up to check the light situation *at the end* of the loop—*not* at the *beginning*. At the end of the work done at one desk, you look up to see if the lights are still on. If so, you will move to the next desk. For this reason, the *do...while* loop is called a *post-test loop*. You test the condition (i.e., the lights) *after* the body of the loop.

```
do
{
execute the body of a loop
while (boolean expression is true);
```

6.5 TWO EXAMPLES FROM A PROGRAMMING PERSPECTIVE

The best way to understand these loops is to start using them. Let's look at an example using each loop. In each example, try to identify the boolean expression used to control the loop. Although there are many situations where any of the three loops could be used, each example uses a loop that works particularly well in the context given.

I.

A good example of the *while loop* is when you use it to get some specific input from the user. Let's use it to verify a password entered by a user at an ATM. This is a good example because we do not want to grant access to an account unless the password is correct. We also want to give the user a chance to correct his password. The *while loop* would test the password initially to make sure the password is correct. Once it is correct, the user would be allowed to leave the loop. The *while loop* would spin as long as the password was incorrect. Here is an algorithm for verifying the user's password. The fragment that executes the algorithm follows.

The Algorithm

1. Ask the user for his password.
2. Check to see if his password is correct.
3. If it is correct, go to Step 6.
4. If it is not correct, ask the user to type it again.
5. Go to Step 2.
6. Allow the user access to his account.

```
string response, password;
.
.
.
//password would be assigned before we get to this
//section.
cout << "Please type your password."<< endl;
cin >> response;
while (response != password)
{
/* The loop will spin when the password is incorrect.
Once it is correct, the user will be "released" from the loop. */
cout << "Your password is incorrect."<<endl;
```

```
cout << "Would you please type it again?"<<endl;
cin >> response;
}
```

II.

Here's an example where we simulate the game-playing scenario we spoke of earlier in the chapter. Let's surround code for a game with a loop that depends on the player's willingness to play the game. We will use the *do...while loop* in this example.

The Algorithm

1. Ask the user if he wants to play the game.
2. Check to see if his answer is *yes.*
3. Play the game.
4. Ask the user if he would like to play again.
5. Check to see whether his answer is *yes.*
6. If answer is *yes,* go back to Step 3; otherwise, stop.

Initially we ask the user for his input and, regardless of whether he says *yes,* we enter the loop that contains the game-playing code.

```
cout << "Would you like to play the game?"<< endl;
cin >> response;
// Even if the user said "no" here, he's going to
// play the game at least once. The do loop always
// spins at least once.

do
{
/*Here is where all the code for the game belongs.
It is probably several pages long. You do not need to know that.
*/

cout << "Would you like to play the game again?"<<endl;
cin >> response;
}
while ( response == "yes" );
```

The user plays the game at least once. Then we ask the user whether he wants to play again. Only if he says *yes* do we execute the *do...loop* a second time. Then he must again say *yes* in order to play the game a second time. You have probably also experienced this kind of loop when you played a game.

```
Would you like to play the game?
yes
```

game is played in here

then again...

```
Would you like to play the game again?
yes
```

and again...

```
Would you like to play the game again?
yes
```

one more time

```
Would you like to play the game again?
no
```

Then it stops. In fact, it would stop for anything that did not look like "*yes*," which encompasses a lot of responses. "No," "NO," "N," "Yes," "YES," "Y," "MAYBE," "okay," and the like would all cause the loop to *stop* becausee the boolean expression (response == "yes") would be *false*. Remember—this is a machine that has *no* intelligence.

6.6 USING A CONDITIONAL LOOP WITH A COUNTER STATEMENT

Any *while loop* or *do...while loop* can be used to simulate the action of a *for loop*. What that means is the *while loop* is set up so that it has all the elements of a *for loop*. That is, it has a control variable, a counter statement, and a boolean condition dependent on the control variable. Unlike the *for loop,* which has these three statements at the beginning of the loop, the control variable is initialized *before* the *while loop* and the counter statement is *within the body* of the while loop. The boolean condition is at the top of its loop, however.

```
int x;
x = 1;
while ( x < 10 )
{
//body of the loop
cout << x << endl;
// counter statement to increase x
x = x + 1;
}
```

As this loop spins, x increases and gets closer to the upper limit, which is 10. Look at the output from executing this loop.

```
1
2
3
4
5
6
7
8
9
```

You will not see "10" on the screen because, once the control variable becomes 10, the boolean expression at the top of the loop is *false* (10 < 10 is false) and we exit the loop. Thus, we never see the "10" on the screen.

When a *while loop* is used with a counter statement, the counter statement acts as a clicker and "counts" the number of times the *while loop* spins. Because the boolean expression is used *to control* the *while loop*, it must depend on the variable used in the counter statement.

We can do a similar example with the *do...while* loop. Recall that the boolean expression will be at the bottom of the loop. Like the *while loop*, we need to set the control variable before we enter the loop and put the counter statement within the body of the loop.

```
int x;
x = 1;
do
{
//body of the loop
cout << x << endl;
// counter statement to increase x
x = x + 1;
}
while ( x < 10 );
```

You will get the same output as you had in the *while loop*. Again, you will not see "10" in the output, because the loop is exited before "10" can be printed on the screen.

```
1
2
3
4
```

```
5
6
7
8
9
```

SUMMARY

We introduced the *loop,* which surrounds one or more programming statements referred to as the *body* of the loop. The loop is used for repeating steps in a program. It is another example of a control statement because the compiler is forced to stay in a loop rather than go to the next line.

Another useful statement is the *counter statement,* which has this syntax: *var* = *var* + 1. Some variable gets its present value plus one. This statement is used to increase the value in a variable and it is also used in the context of loops. When a counter statement is used in a loop, it allows the variable to become bigger (or, conversely, smaller) so that the boolean expression controlling that loop can be altered.

We examined two kinds of loops—*fixed iterative* and *conditional* loops. The fixed iterative (repetitive) loop executes a known or fixed number of times. The conditional loop will spin as long as a boolean expression is true.

The *for loop* is a *fixed iterative* loop. It spins a known number of times. The loop has three parts besides the body of the loop: a *statement initializing* a control variable (assigning it a first value), a *boolean expression,* and a *counter statement* using the control variable.

Two conditional loops are the *while loop* and the *do...while* loop. The *while loop* is an example of a *pre-test* loop. A boolean expression is tested (examined to see if it is true) *before* the loop is entered. The *do...while loop* is an example of a *post-test* loop. The condition is tested *after* the loop is executed. Both the *while loop* and the *do...while loop* can be used to replace a *for loop* by inserting a counter statement into the body of either loop.

ANSWERS TO ODD-NUMBERED EXERCISES

6.3

1. a)
3. a)
5. g = g – 1

CHAPTER

7

Function Calls: That's What It's All About. Get Somebody Else to Do the Work.

IN THIS CHAPTER

- What Is a Function?
- What Needs to Be Sent into a Function (Helpers or Parameters)
- What Comes out of a Function—Return Values
- Two Kinds of Parameters: Value/Copy Parameters and Variable/Reference Parameters
- How to Call a Function
- Inside the Main When You Make a Call
- Inside the Function When You Get "Called"

7.1 WHAT IS A FUNCTION AND WHY USE ONE?

In this chapter, we define functions and look at all the programming concepts addressed by them. Functions are *separate blocks of code* that appear before or after the main section we learned about in Chapter 4. A program will appear separated into blocks (Figure 7.1).

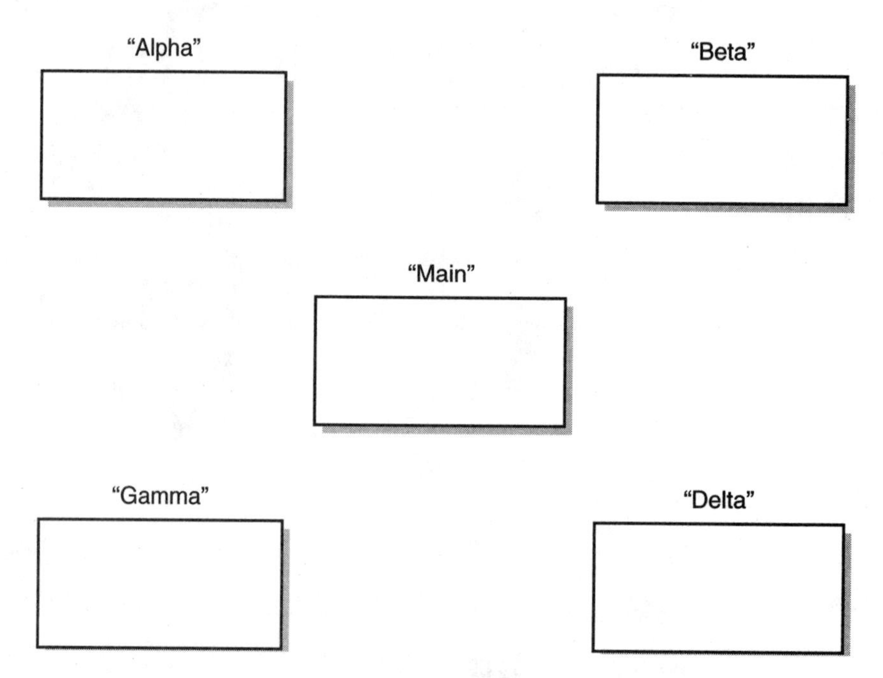

FIGURE 7.1 A schematic of a program shows each separate block with its name above. Each block represents a separate area with its own code.

Functions are a way of *organizing* a program into separate blocks to accomplish certain tasks. Imagine that you were writing a program to simulate all the tasks that a calculator could do on a number or pair of numbers. This program would be ideal for using functions because there are so many separate tasks that need to be programmed.

A calculator program would do many different things. Think of all the separate tasks that such a program would execute, for example:

- Adding two numbers
- Subtracting two numbers
- Multiplying two numbers

- Dividing two numbers
- Converting an answer to scientific notation
- Changing from degree to radian measure when working with angles

Now we could have a large program with a separate function for each of the tasks we listed (Figure 7.2).

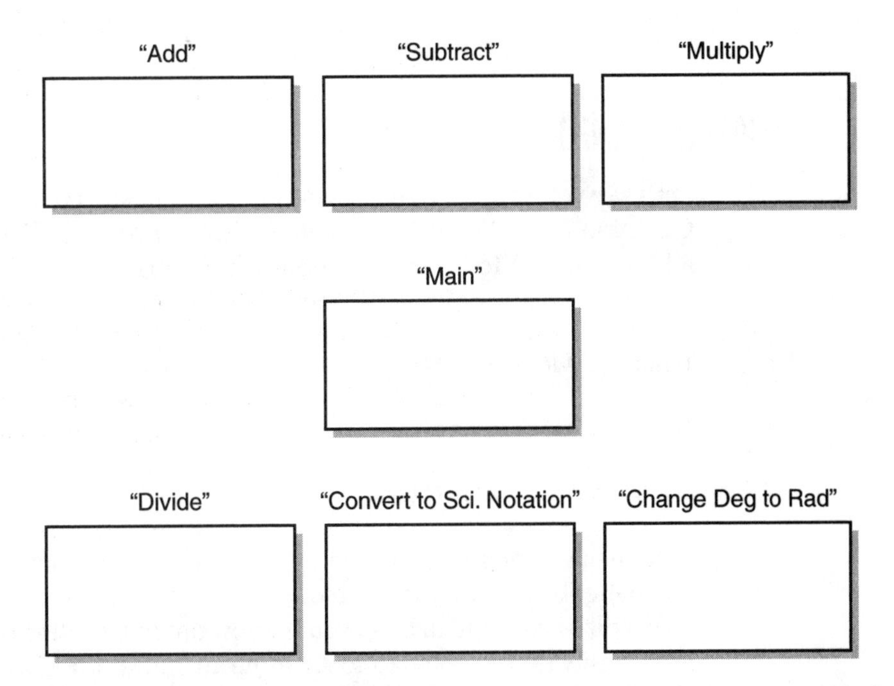

FIGURE 7.2 A schematic of a program is shown with each task as a separate block that would contain code to execute that task.

Using functions on a complex problem allows the programmer to separate her code into separate blocks. By organizing code in this manner, it will be easier for someone else to read and understand her code. Functions are usually blocked off in this manner—they appear outside of the main section and they are *called* by the main when needed to be executed.

Another reason for using a function is to block off code that you know you will use *repeatedly*. Let's say you write a program to maintain a savings account. The program should print out balances after each transaction. So, ideally, you would write a function to print the balance in the savings account. Every time you do something to the account—like make a deposit or a withdrawal— your program calls one of these functions to do the task. Here are three functions that would be included in your program.

```
function  Print_Balance

function  Make_A_Deposit

function  Make_a_Withdrawal
```

ANOTHER EXAMPLE

Let's say you want to write a program that will keep track of all the music CD's you have. You might want to write a program that allows you to add a new CD to your collection, alphabetizes your list of CD's, and prints all the titles. Here are functions for each of these tasks

```
function Add_New_Title

function Alphabetize_List

function Print_All_Titles
```

By using functions, you are organizing the program so that a reader will better be able to understand what you are doing. Furthermore, your code will be easier to debug because you can find your error in a function more easily than if you had to examine an entire program.

7.2 WHAT IS A FUNCTION AND WHAT DOES IT DO?

A *function* in mathematics is a *set of directions to manipulate a variable*. It is really a collection of operations on a variable. Think of what an adding function would do to two numbers like 3 and 5—3 + 5 produces 8. We could use other operators with four numbers—3, 2, 18, and 7—to produce a more complex result—3 + 2 * 18 – 7 produces 32. In each case when using a function, think of the numbers that *come into* the function and the answer that *goes out* of the function.

In computer programming terms, *a function is a separate body of code that performs some task*. Think of a function as a machine that has certain instructions to perform—usually on a variable or variables that are being *sent into* it. Along with doing the instructions posted inside of it, a function "machine" will probably have some values or variables that *come out* of it (Figure 7.3A and B).

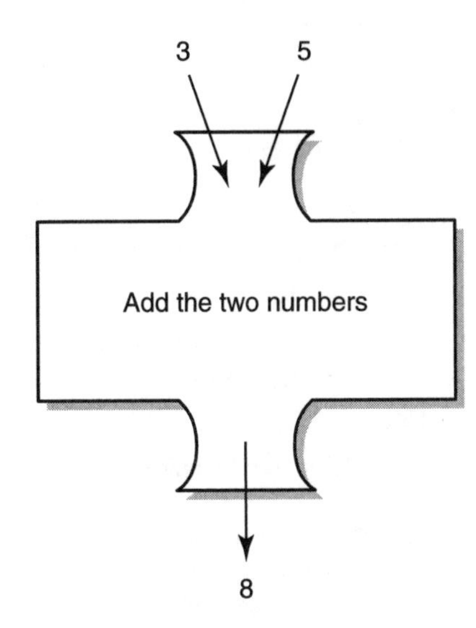

FIGURE 7.3A A function is shown as a "sum" machine where two numbers are sent into it and a sum comes out of it.

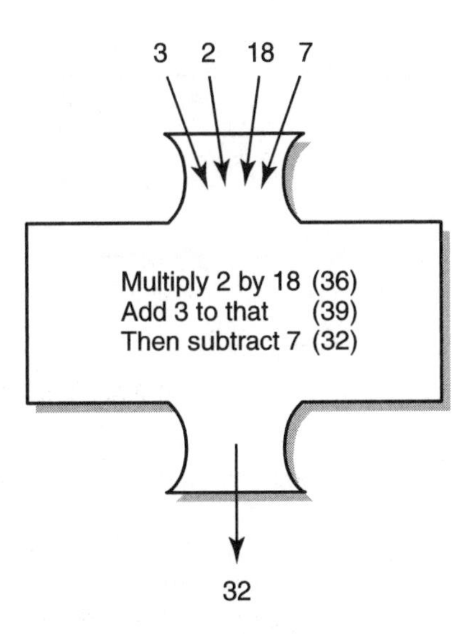

FIGURE 7.3B Another function is shown, where four integers are sent into it and a result comes out of it after the operations are performed on them.

FIRST EXAMPLE

A function finds the square of a number. Notice that in the "Squaring" function, the number to be squared is sent into it and the result comes out of it (Figure 7.4).

SECOND EXAMPLE

A function prints a message *the number of times specified*. The *number* of times specified will be *sent into* the function and there will be *no values coming out* of the function once it has finished (Figure 7.5).

In our last example, we consider a function that has nothing going into it and nothing coming out of it. It is simply a function that does some task for the programmer. An icon or picture drawn by a computer is a separate task that is ideal for a function to do. The programmer does not have to write the code herself to draw the picture she wants. Look at function "Draw Circle" (Figure 7.6).

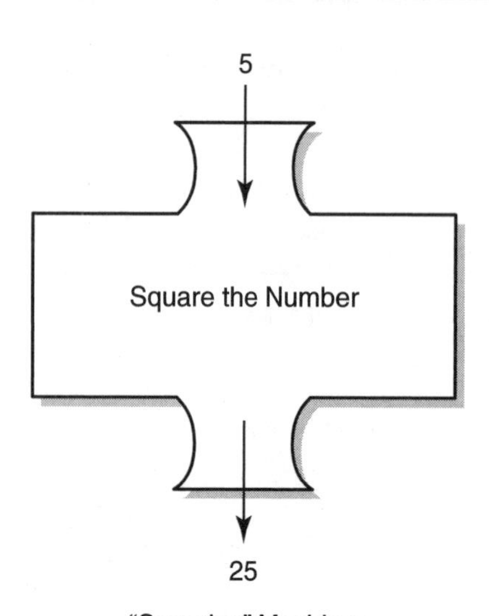

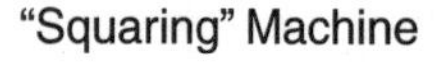

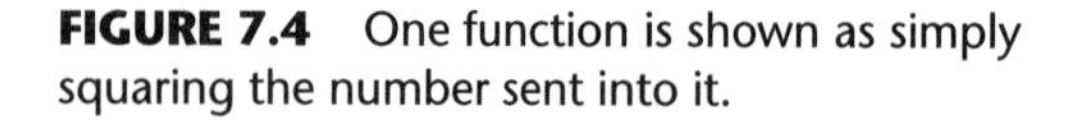
FIGURE 7.4 One function is shown as simply squaring the number sent into it.

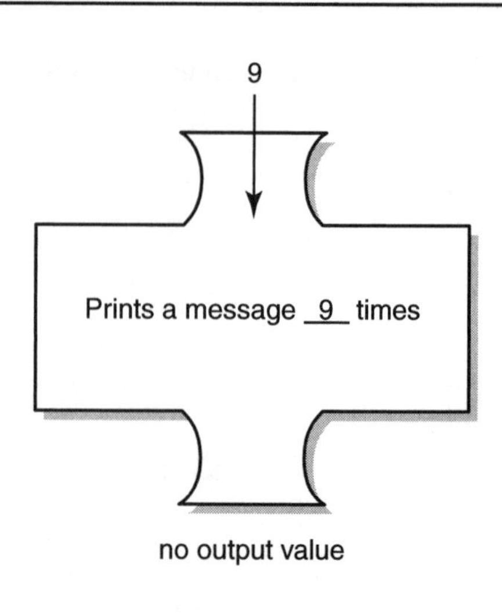

"Printing" Machine

FIGURE 7.5 The "Printing" function is shown with a number coming into it so that a message can be printed that number of times. Once the function is complete, there is no value coming out of the function.

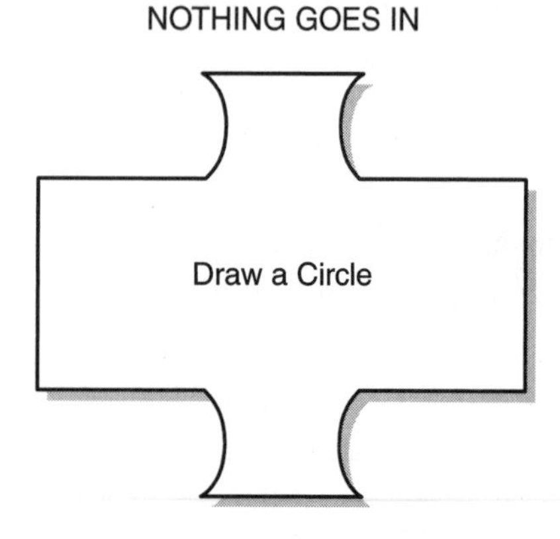

NOTHING COMES OUT

FIGURE 7.6 The "Draw Circle" function is shown with no value going into it and no value coming out of it.

IMPORTANT PARTS OF A FUNCTION

In the previous examples, we spoke of *values coming into* the functions and *results coming out* of them. This is the case most of the time. In the "Printing" function, a value came into the function (9) but no value came out of the function because our task was simply to print a message. Compare this function with the "Squaring" function. The "Squaring" function accepts a number and then executes code to produce the square of the number.

Values (usually stored in variables) *both coming into and going out of the function* are *two key concepts* of *function use*. It is important that we stop to address all the issues that surround these aspects.

7.3 HOW FUNCTIONS INTERACT WITH THE MAIN

When a programmer wishes to go to a function to complete some task, she must write a command to *call* the function. Before we examine all the details associated with calling functions, let's examine some analogies from everyday life. Calling a function is like calling a friend to ask her to help you complete a big project—like painting the interior of a house. You call your friend, Ingrid, to ask her to paint one room and give her what she needs—the paint, sandpaper, and some brushes—to do you the favor. When she is done, she tells you she finished it. This is an analogy of interacting with a function.

In another example, your VCR is broken and you call a friend, Jack, who is really good at fixing broken VCRs. You call Jack on the phone to ask him if he can help you. He says "*yes,*" but says he'll need some tools—a cassette for testing the machine and a replacement part for the broken part. When Jack is done repairing the machine, he calls you back to say he has finished and he returns the VCR to you.

Let's review our two examples to see what happened in each situation. By analyzing each case, both from a general and specific point of view, we will better understand how a function calls work.

General	*Specific*
Call your friend.	Call Ingrid.
Give your friend something.	Sandpaper, brushes, and paint.
You are called back.	Ingrid calls you to say she finished.
You're given something (yes/no).	No.

General	*Specific*
Call your friend.	Call Jack.
Give your friend something.	Broken VCR, cassette, and replacement part.

You are called back.	Jack calls you to say he finished.
You're given something (yes/no).	Yes. Jack gives you back the VCR.

When Ingrid or Jack did us a favor, they first got something from us to do the work. Then, they called us back to say they had finished. In Jack's case, he returned the VCR to us. In Ingrid's case, she didn't return anything to us, but she accomplished the task—painting the room.

Functions work *almost* the same way. The programmer writes a command that is a "call." In the call, she includes variables that the function needs to complete its work. The function executes its code and after that the *compiler leaves the function to return to the place that called it, in order to continue executing the rest of the program.* At that point, the compiler *may* return something to that place—only if necessary.

Let's take four examples of functions that we discussed in Section 7.2 and give them names so that we can "call" them— "Sum," "Fun_With_Nums," "Print, " and "Draw Circle" functions.

General	*Specific*
Call the function.	Call "Sum."
Give the function something.	Two integers.
Compiler "goes back."	Compiler moves from "Sum" to "main."
A value is returned (yes/no).	Yes. "Sum" returns an integer.

General	*Specific*
Call the function.	Call "Fun_With_Nums."
Give the function something.	Four integers.
Compiler "goes back."	Compiler moves from "Fun_With_Nums" to "main."
A value is returned (yes/no).	Yes. "Fun_With_Nums" returns an integer.

General	*Specific*
Call the function.	Call "Print."
Give the function something.	An integer.
Compiler "goes back."	"Print" calls "main" when done.
A value is returned (yes/no).	No. "Print" returns nothing.

General	*Specific*
Call the function.	Call "Draw Circle."
Give the function something.	Nothing.
Compiler "goes back."	Compiler moves from "Draw Circle" to "main."
A value is returned (yes/no).	No. "Draw Circle" returns nothing.

Notice that in the last example—the call to "Draw Circle"— *nothing was sent in* and *nothing came out* of the function.

Both Ingrid and Jack called us back to say they had completed the favor. In the first example, Ingrid simply called us to tell us she had finished the favor. In the second example, Jack called us to tell us he had finished the work and was *returning something to us*—the VCR.

Functions behave the same way. The compiler "will leave" the function when it finishes executing its code and go back to the next line of code following the "call." We need to know whether a function returns something after executing its code. This is called *returning* a value. Later in this chapter, we will look at the specifics of how to call a function. For now, it is only important that you understand what is meant by calling a function.

```
int main  (  )
{
.
.
.
//insert call to function here
//compiler will return to this line.
.
.
.
}
```

After executing a function, the compiler will return to the part of the program where the call was made.

FUNCTION NAME AND PARAMETER LIST

In most languages, a function has a *name* and a *parameter list*. The *name* is used whenever the function is *called*. Once the function is called, the compiler goes to the function to execute it. Here again, we see another example of controlling the compiler. The compiler leaves the line it is on and "goes" to the code of the function to execute it.

The *parameter list* is a list of *all the variables and their types sent into the function*. It appears to the right of the function name and gives us information about what kinds of values we can expect to see sent into this function.

HINT!

Think of the word "parameter" in English. It usually means a *boundary* or *restriction*. In the language of computer programming, we are restricting *how* a function will operate. What is sent into a function through the parameter list determines what the function will produce most of the time. So the parameter list is like the boundaries of the function.

RETURN VALUES

Functions usually return values back to the main—the part of the program where the call was made. They produce a result in the form of some specific *type* of variable and *that result is sent back* to the main function after the "called" function has been executed.

Let's examine some of the previous examples to see which results were sent back. In the first example, where we added 3 + 5 to get 8, *8* (an integer) was the result. In the second example, which involved more operators, 3 + 2 * 18 – 7, *32* (also an integer) was the result. In the squaring function, where 5 was sent into the function, an integer was the result. In the last function we examined—the print message function—there was *no result*—only the actions taken by the function.

Let's start by reviewing what each function did:

Function	*What the Function Did*
Sum	Added two numbers
Fun_With_Nums	Took 4 numbers and performed three operations on them

Square	Squared a number
Print	Printed a message

Now we need to check what type of result is produced. If a function has no result, we use the term "*void*" to indicate that nothing is produced.

Function	Did it Produce a Result?	What Type?
Sum	Yes	A number
Fun_With_Nums	Yes	A number
Square	Yes	A number
Print	No	Void

Now we need to check to see whether the number produced is an integer or a real. Consider these examples with each function and the values that result from sending different numbers into each function.

Example	*Send In These Numbers*	*Result*	*Type*
Sum	5, 12		
What the function does: (adds the two values)			
5 + 12	⇒	17	⇒ integer
Sum		14, 17.2	
What the function does: (adds the two values)			
14 + 17.2	⇒	31.2	⇒ real
Fun_With_Nums	6, 3, 5, 8		
What the function does:			
(adds, multiplies, and subtracts—			
multiplication will be first!)			
(6 + 3 * 5 – 8)	⇒	13	integer
Fun_With_Nums	12.4, 5, 8, 4.2		
What the function does:			
(adds, multiplies, and subtracts—			

multiplication will be first.) (12.4 + 5 * 8 – 4.2) 12.4 + 40 – 4.2 52.4 – 4.2	⇒	48.2	⇒ real
Square What the function does: (squares a number) 6 * 6	6 ⇒	36	⇒ integer
Square What the function does: (squares a number) 4.2 * 4.2	4.2 ⇒	17.64	⇒ real
Print What the function does: (prints a message 5 times)	5	void	void

The value *that is returned* after the function is executed will *depend on what types are sent into the function and what the function does to those types.* Keep in mind that sometimes the same type "comes out" as "went in." You also want to remember that there are functions that *do not produce* anything to be sent back to the main section. In this case, we use the term "*void*" to indicate that nothing is coming back to the main function.

7.4 HOW TO WRITE A FUNCTION HEADING

A *function heading* is a line of code that tells the compiler all the important information it needs to know about the function. There are *three* parts to the *function heading*. First, the *return type* is listed to tell the compiler that this function will produce an integer, for example, when completed. The term *void* will signal the compiler that *nothing* will be returned by the function. The next part of the function heading is the function *name*. If you are the compiler and you need to call on a function to do work, you need to *call it by name*. The third part of the function heading is the parameter list—the list of variables and their types sent into the function.

Return Type	Function Name	Parameter List
int	*Sum*	(int x , int y)
double	*Fun_With_Nums*	(double a, int b, int c, int d)
double	*Square*	(double x)
void	*Print*	(int num_times)

Combining return types with the name of the function and its parameter list creates function headings like the following:

```
int Sum ( int x , int y )
double  Fun_With_Nums (double a, int b, int c, int d)
double  Square (double x)
void   Print ( int  num_times)
```

7.5 PARAMETERS: TWO DIFFERENT TYPES

Parameters (variables sent into functions) are boundaries or restrictions on a function's behavior. When you examine a parameter list, you learn a lot about how a function behaves. Does the function need one integer and one real? Does it require two integers? Or maybe it needs three doubles and one string. By reading a parameter list, you get a sense of the restrictions on the function's work. A parameter list tells us the requirements of the function much like the analogy of what our friends needed when they did us a favor. Ingrid needed the materials to paint the room and Jack needed a replacement part for the broken VCR.

In addition to the type of variable being an important aspect of a parameter sent to a function, there is one other aspect of a parameter list. That is, how the variable is sent into the function. Is it a *copy of itself* or is it being sent itself? This question gives rise to the *two different types* of parameters. As we examine these two types, notice how each variable gets sent into the function. We will learn about two different kinds of parameters: the *value (copy)* parameter and the *variable (reference)* parameter.

VALUE (COPY) PARAMETERS

When variables are sent into functions, they can be sent in two different ways. One way is that the function sees that a variable is coming into it and it generates its own *copy* of the variable that is coming in without altering the original variable.

Think of it this way. We will go back to our analogy of Ingrid painting a room for us. Instead of Ingrid taking from us the sandpaper, paint, and brushes that we want her to use, she talks to us on the phone to find out *exactly* what type and color of paint we want to use, what kind of sandpaper—coarse or fine—as well as the type and size of brush. Then she buys *her own exact copies* of what we would have given her. We keep our brushes, sandpaper, and paint and *she uses* the ones she bought.

She copied everything we provided and used her own items. The only advantage of her buying her own items is that maybe she didn't have to carry our items over to the house. She just called us on the phone to check all the parameters—the required materials—for the task.

Value (Copy) parameters behave the same way. They get all *the values* of the original variables but make their own copies of those variables for the function with the exact same values inside of them (Figure 7.7).

HINT!

A function can do anything to these *value parameters* because the original variables of the main will not be touched. This aspect is one of the main reasons that programmers use *value parameters.*

VARIABLE (REFERENCE) PARAMETERS

Variable parameters, also called reference parameters, are the second type of parameter. Here the parameters *do not create* new *copies* of the original variables, but, instead, *refer back to* the original variables. In a sense, the function has *full access* to the *original variables* used in the main function. The reason these parameters are called "variable" is because the word "variable" (in English usage) means "changeable." A variable parameter has the power to change or alter an original variable back in the main function.

The intention of the programmer when she uses a *variable parameter* is that she wants to let the function manipulate the original variable. The function "works on" that variable and alters it (Figure 7.8).

HINT!

With the *variable parameter,* the function gets its hands on the original variable and not a copy of it. This aspect of parameter passing can be useful if the variable *takes up a lot of memory* and you *do not want to waste more memory* by generating a copy of it for the function. You decide to use the original.

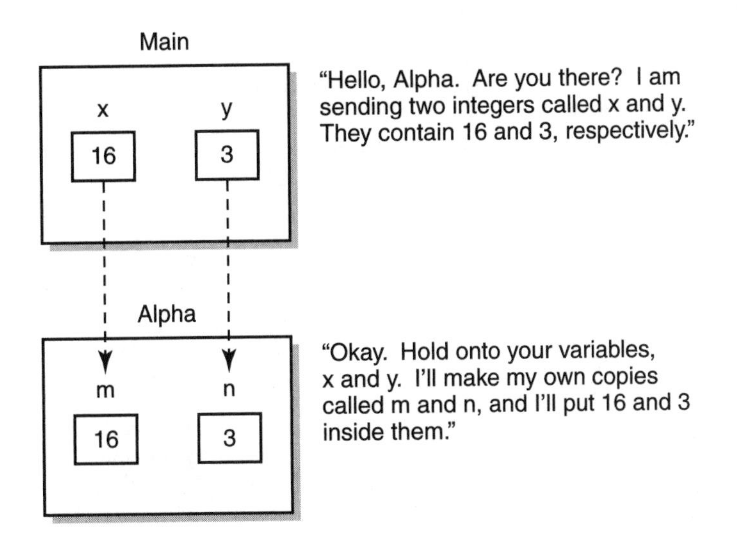
Main
x
y
16
3
"Hello, Alpha. Are you there? I am sending two integers called x and y. They contain 16 and 3, respectively."
Alpha
m
n
16
3
"Okay. Hold onto your variables, x and y. I'll make my own copies called m and n, and I'll put 16 and 3 inside them."

FIGURE 7.7 A conversation is shown between a "main" function and a function called "Alpha." "Alpha" does not want to be responsible for the "main's" variables (x and y), so it generates its own copies (m and n) of those variables.

The reason it is called a *reference parameter* is because the parameter *refers back to* the original variable—not a copy of it. A reference book is a useful comparison because, in a reference library, you cannot walk off with this book—you can only look at it there.

The difference with *a variable parameter* is that it *can do whatever it wants to the original variable*—including *change its value—from inside the function*. In our comparison with the reference book, the variable parameter has more privileges than you do. When you use a reference book, you cannot write in it or alter it while a variable parameter can do whatever it wants to the variable it references (the variables in the main).

THE SYMBOL FOR A VARIABLE (REFERENCE) PARAMETER

Variable parameters are designated by a *symbol used in front of the name of the parameter* in the parameter list of the function heading. In the programming language of C++, the ampersand symbol, '&,' is the symbol used. The symbol (&) is the only way to indicate that a *parameter will refer back to* an original variable (in the main) *rather than generate a copy* of that original variable.

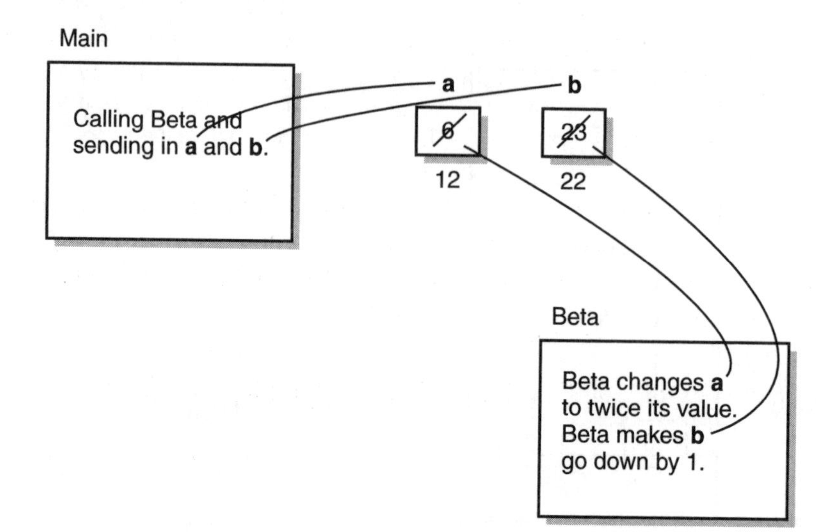

FIGURE 7.8 A "main" function is shown with its own variables, *a* and *b*, and their values. Another function "Beta" is shown affecting the values in those variables. "Beta" doubles *a's* value and makes *b's* value drop by 1. Since *a* started out as 6, it ends up being 12. Since *b* started out as 23, it ends up being 22.

EXAMPLE

`int`	`Alpha`	`( int & x, int & y );`
return type	function name	var type with reference symbol and parameter name

HINT!

If there is *no symbol in front of a parameter name,* then the parameter is a *value parameter* and it will be a copy of an original variable.

7.6 HOW TO CALL A FUNCTION

When you call a function from the main function, you direct the compiler to go to the function bringing with it any variables you are sending into the function. When you write a call, you use the syntax of the function heading as a guide. For example, if a function *heading* has a return type and a parameter list containing 2 variables, it might look like this:

```
double Find_Average ( int first_num, second_num);
```

A function *call* will look exactly like the heading *without* any of the *variable types mentioned.* Let's practice developing the syntax for a *call—the command to get the compiler to leave where it is and go to a function.* We'll do this in stages because there is a lot of detail in this process.

WHERE THE CALLING BEGINS: INSIDE THE MAIN FUNCTION

The first part of the call is to use the function name:

```
Find_Average
```

The next step is to follow the restrictions of the parameter list by examining the parameters that *Find_Average* has. That is, we need two integers from the main to match those in the parameter list.

```
Find_Average (              ,              )
                 ⇑              ⇑
             an integer     an integer
```

But wait! Now comes the tricky part. *First_num* and *second_num* are actually variables that *belong to* the function *Find_Average* because those were the names used in the heading. They do not belong to the main function. So the main needs to have its *own* variables whose values will be passed into the *Find_Average* function. Let's declare some variables of the *exact same type* given in the parameter list. We need two integers. Let's declare them and assign them values.

```
int  x, y ;
x = 5;
y = 7;
```

Now we can insert them into *the call* itself. (We're almost done.)

```
int  x, y ;
x = 5;
y = 7;
Find_Average ( x, y);  // this is an incomplete call!
```

Remember that *Find_Average* (which uses *value* parameters—there is no '&' symbol!) *will copy the values* of x and y, but, ultimately, leave those variables alone. *Find_Average* looks at x and y and *generates copies* of those variables and *calls them* by its *own designated names* of *first_num* and *second_num*.

The last stage of developing a call is to handle the return type. When *Find_Average* is done with its work, it will *send back* a variable of type *double* to the main function, which called it. So we need to be ready to "catch" what it sends back. We declare a double variable and use it in an assignment statement like this:

```
int  x, y ;
double avg;
x = 5;
y = 7;

avg = Find_Average ( x, y); // the call is now complete
```

Now we can analyze the syntax of a correct call.

avg	=	Find_Average	(x, y);
return type	assignment operator	function name	variables from the main that will be copied by function

THE RECEIVING END: INSIDE THE CALLED FUNCTION, FIND_AVERAGE

Once the call has been made, the compiler enters the function but now is doing its work—that is, executing code with the copies of the variables and their new names— here, *first_num* and *second_num*.

```
double Find_Average ( int first_num, int second_num);
{

// developing an average for two things
//  add the two variables like this:
//   first_num  +  second_num
// then divide that answer by 2 for the average.

// (first_num  +  second_num ) / 2

// I need parentheses to do addition first.
}
```

We're almost done! We just need to handle the part about returning values. *We use a return statement as the last part of the function.* When the function has finished its work, it needs to *return a value* that is of the *double* type, as specified in the *heading*. (Remember when Jack *returned* the VCR after he had finished repairing it?) We do the same in a function. We *return the value.* Before we do that, we will declare a variable of the *double* type.

```
double Find_Average ( int first_num, int second_num)
{
double the_average;

the_average =  (first_num  +  second_num ) / 2 ;
// I did the same thing I did in the "main."
// I assigned the result to the variable.
// Now I can return it.

return the_average;


}
```

In order to avoid confusion about the different variable names, let's look at it this way. Imagine two rooms—one called the "main" room and another called the "Find_Average" room. Each room has its own variables. These variables can't leave the room they're in—they are only recognized in their respective rooms. When the main makes a call to a function, it uses its own variables to do this. On the other end, the function receives the call and "catches" these variables *with its own* variables of the *exact same type.* Variables in calls must match variables in headings *type for type* (Figure 7.9).

Each variable is only recognized in its room so we need to have different names for the call and the heading.

The "Main" Room	*The "Find_Average" Room*
`x , y`	`first_num, second_num`
`avg`	`the_average`

SUMMARY

In this chapter, we covered the topic of functions. Functions provide a way of organizing code in a program so that any reader of the program will be able to understand the program easily. Calling a function to do a task (like printing a chart) whenever you need it done is easier than repeating a block of code in your program. Your program will be shorter and more organized through the use of functions.

We gave some examples of different kinds of functions—both those that produce a value as well as those that execute a task such as printing some message repeatedly. Functions have very specific syntax (grammar) requirements. You need to write a function heading that includes

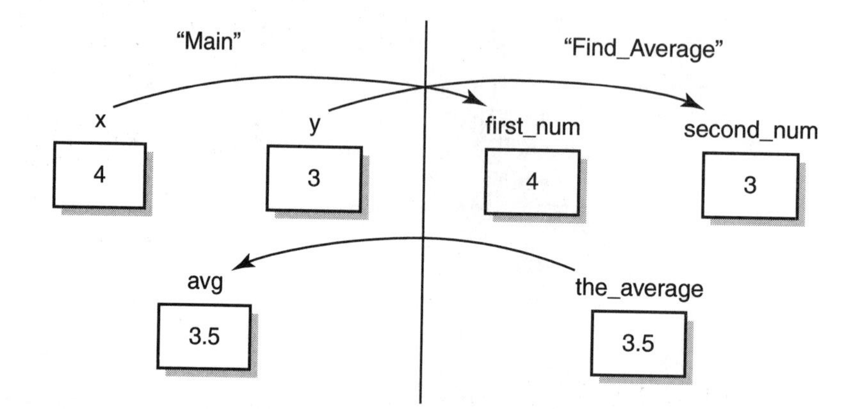

FIGURE 7.9 Variables as they appear in their own rooms. Each variable has its own box to show that it contains a certain value. In this example, we will send the numbers *4* and *3* into the function and get back the value *3.5* after the average of *4* and *3* is computed. Note the direction of the arrows to show which variable "sent" its value back to the other.

a function name and a parameter list, and a return variable type, if necessary.

Parameters are the names of variables that the function needs from the main function (the calling place) of the program. There are two types of parameters: *value* (also called *copy*) parameters and *variable* (also known as *reference* parameters). The variable parameters are designated by the '&' symbol in front of them.

When you *call* a function, it is important to match the variables from the main *type for type* with those in the function heading. It is also a good practice to use *different* variable names in the function heading so there will be no confusion between the variables used in the call (in the main function) and those used in the heading of the function.

CHAPTER

8

Using Functions in Graphics

IN THIS CHAPTER

- Define Graphics
- Vector Graphics
- How Do Graphics Work?
- Drawing Lines and Points
- Examples of Functions That Generate Graphics
- Typical Graphic Moves
- Making Some Drawing Disappear
- Moving Some Drawing
- "Delaying" the Computer
- How to Call a Function

8.1 GRAPHICS: HOW DID YOU DO THAT?

The topic of *graphics* brings to mind anything from complicated 3-D games such as *Doom* and *Final Fantasy* to the antiquated graphics used on highway notification signs.

The subject of *graphics*, creating and displaying images on a computer, will continue to evolve as languages make greater use of the digital images travelling over the Internet. Most of the work will be done for you. The programmer will only need to position the image and not have to draw it himself. No doubt using graphics will be easier and easier if a student only has to import an image and then position it to his liking.

In this chapter, we will look at some of the technique involved in creating and displaying images. By examining older graphics methods, you will feel confident that you will be able to handle whatever newer languages offer in the way of graphics.

VECTOR GRAPHICS

The word vector implies magnitude and direction. Think of a vector as an arrow that has a length and a direction. (Is the arrow pointing down or up or sideways?) Imagine drawing a picture with the following restrictions on your drawing. First of all, you can only draw an image with straight lines. What could we draw if someone restricted us in this manner (Figure 8.1)?

What if someone forced us to draw our pictures sequentially? We would draw one arrow and then the next arrow would begin where the previous arrow left off. The third arrow would begin where the second arrow left off, and so on (Figure 8.2).

FIGURE 8.1 A house is drawn that has only one door and two square windows. Grass on the front lawn is depicted through lines.

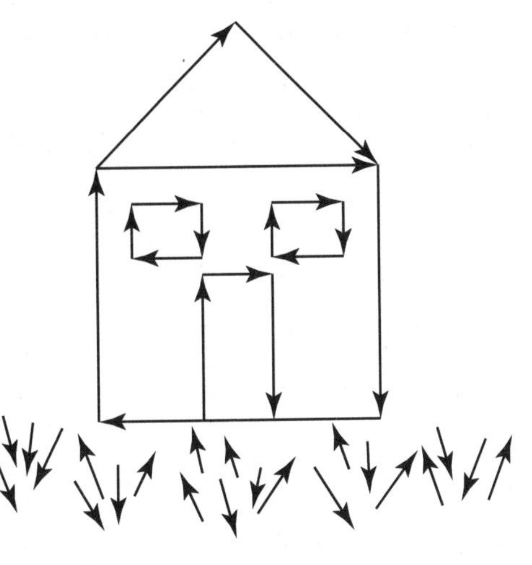

FIGURE 8.2 A house is drawn as a sequence of arrows joined head to tail to create the final image of a house.

Creating an image from line segments that touch each other is an example of vector graphics. In reality you do not need to have each new line touching the previous one. However, it is helpful to remember where lines begin and end.

8.2 DRAWING LINES AND POINTS

To draw lines and points, you first need to understand that you are drawing on a "piece of paper"— the screen—that has been marked as a grid that has rows and columns. The rows are numbered, beginning in the top left at Row 1 and Column 1, and continuing down to Rows 2, 3, 4, and so on— to Row 150, for example. The columns start at the top left and continue to the right for columns 2, 3, 4, and so on—all the way to column 250, for example (Figure 8.3).

The actual number of rows and columns will depend on your computer screen. Once you understand how the screen has been delineated into these blocks, then you can start to draw. The way to draw is to decide *where* you would like to start drawing a line.

Imagine that you are holding a pen in your hand and you want to put the pen down onto the "paper." Where do you want to put it? Let's start by putting the pen onto the block marked 5, 40. That means that we have

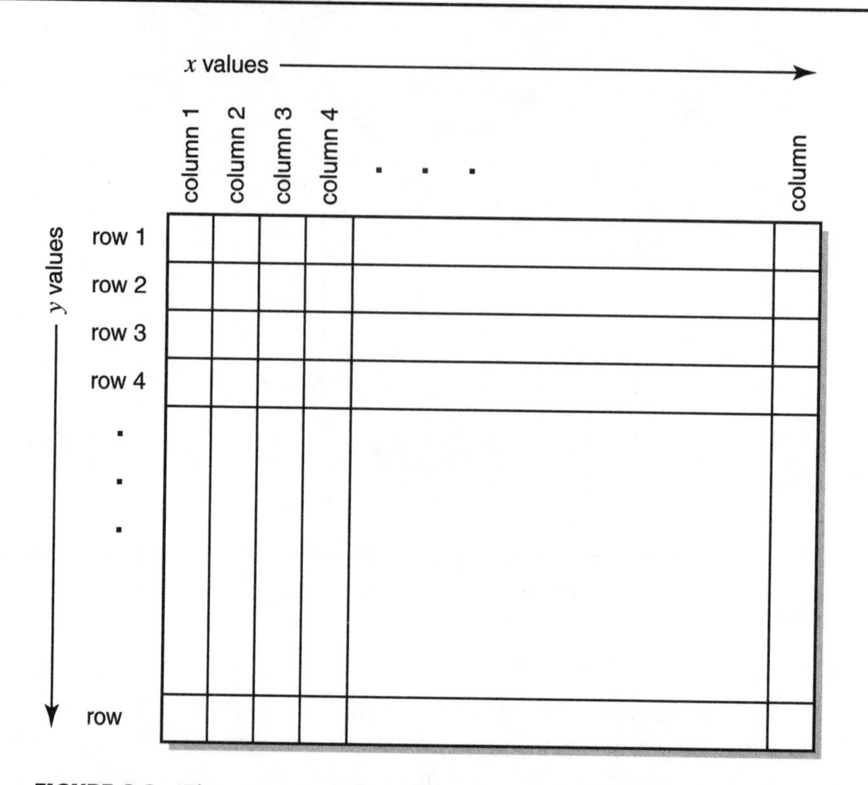

FIGURE 8.3 The screen is shown with an imaginary grid imposed over it to show the positions on the screen.

put the pen down onto the "paper" at Column 5 and Row 40. We haven't drawn anything until we call on a function to help us draw something (Figure 8.4).

Let's look at some headings of functions that will be useful for our drawing. Our initial position (5, 40) is a location on the grid. Let's call those two integers taken together, a *point*. The first function is called *MoveTo* and we will use it to move to a different point from the one we are currently on (5, 40). Because it is a function, we need to examine its parameters so that we can call it properly. The value parameters of the function *MoveTo* are *two integers*—one for the *row* and *column* positions of the new point where we intend to move.

EXAMPLE

```
MoveTo (50,45);  // this call will cause us to move to column 50
// and row 45 on the grid
```

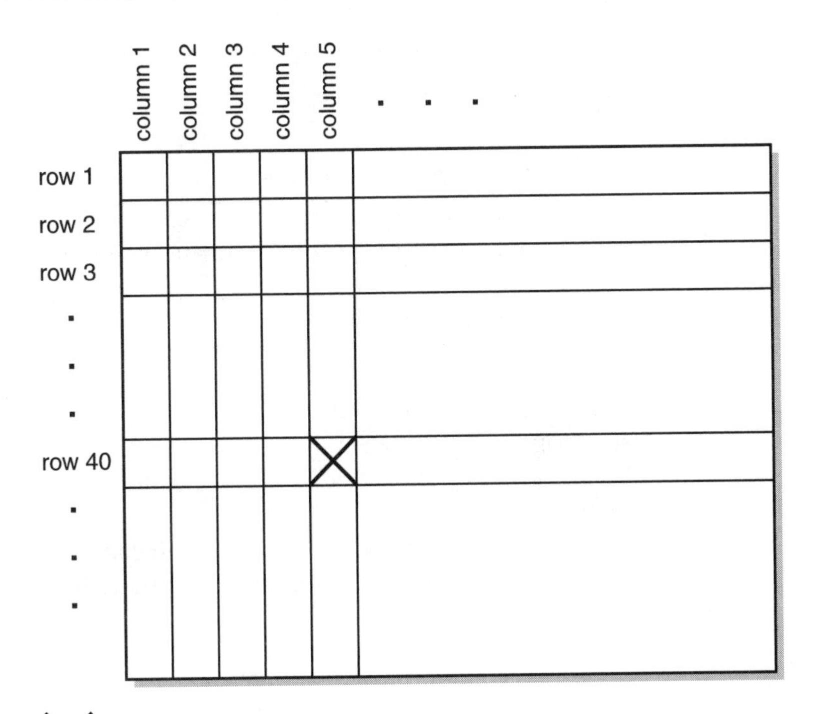

FIGURE 8.4 An "x" is shown on position 5, 40 on the superimposed grid on the screen. (You cannot see the grid on the screen.)

```
MoveTo ( 5,5); // this call will cause us to move back up
// to the upper left corner of the screen.
```

HINT!

Recall that nothing is drawn by the *MoveTo* function. We are simply moving to a new position on the screen.

The second function, *LineTo,* also takes two integer parameters to represent the two coordinates (the column and the row) of the *point* on our grid where you wish to draw a line. *LineTo* will draw a line from your present position on the grid to the new position sent into the function.

EXAMPLE

```
LineTo (25, 20);
```

In the example, a line will be drawn from our present position (5, 5) to the new position (25, 20), as specified by the call to the *LineTo* function.

Here are the headings for each of the functions.

```
void MoveTo (int x, int y)

void LineTo (int x, int y)
```

Both functions have the word *void* in the return type position. This indicates that *no values* will be *sent back* to the calling place after the functions have been executed.

So let's practice using these functions so you can see how they work. The first thing we will do is draw a huge capital letter 'M.' We'll draw this by using the following algorithm.

Algorithm for drawing an M.

1. Move to the lower left corner of the letter.
2. Draw a line straight up. |
3. From that position, draw a diagonal line down to the middle of the *M.*
4. From there, draw another diagonal line up to the top right corner of the letter.
5. Draw a line straight down to finish the letter.

The algorithm involves moving to where we want to position the *M* and then drawing the four line segments that comprise the *M*. The difficult part about making the letter will be judging where the two diagonal lines should meet in the middle of the letter.

M

Find the middle of this letter.

The column should be halfway between the left and right columns.

Step 1. Move to 10, 45 position.
Step 2. Draw a line up to 10, 25.

That means that our line is 20 units long. That is, the line stretches from Row 45 up to Row 25. (Remember we are moving in a direction as we draw this letter. We are trying to draw in sequence and not just any way that we want. Otherwise, we would constantly have to reposition ourselves by calling the *MoveTo* function.)

Step 3. Draw a diagonal line down to Row 40 (not as deep as Row 45) and at Column 30.

Step 4. Draw a diagonal line up to Row 25 (same height as left vertical line) but at Column 50.

Notice that these last two steps position the top tips of the *M*, 40 units apart. (The left tip is at Column 10 and the right tip is at Column 50.) The middle point of the letter (let's call it the vertex) is at Column 30. By adding or subtracting 20 columns from this vertex we get *symmetry*—a natural mirror like quality where one side of the letter is a mirror image of the other.

Step 5. Draw a line from the right tip down to the bottom. (From the Point 50, 25 to 50, 45)

Let's look at each of these steps as a connection of points on a surface. Consider that you map the points first then afterwards you connect them to create the letter *M* (Figure 8.5).

Now let's list the sequence of steps to draw the letter *M* that would be used in a program. The first part of the program is to include the library

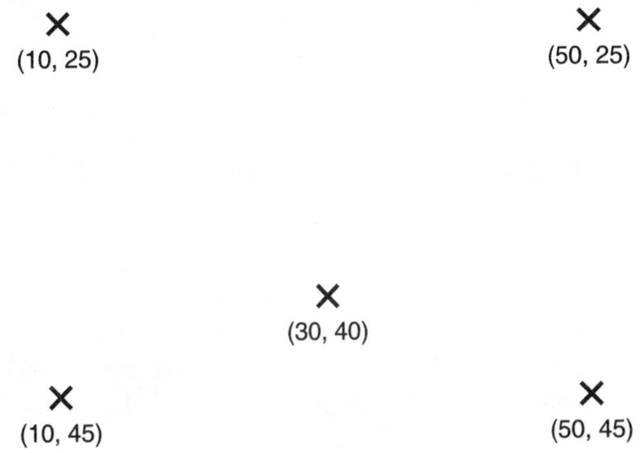

×
(10, 25)

×
(50, 25)

×
(30, 40)

×
(10, 45)

×
(50, 45)

FIGURE 8.5 The outline of the letter *M* is shown as a group of unconnected points on a surface.

(group of files) that contains these functions. If you recall, we have already seen the iostream library (that has the functions that facilitate showing output and receiving input) and the string library, which has the functions that manipulate strings. A graphics library will include these functions that I have mentioned. So let's examine a small program to draw the letter *M*. The program is not complete because some of the initial commands to open a window on which to draw will vary from machine to machine.

```
#include iostream.h
#include graphics.h

int main ()
{.
 .
 .

MoveTo (10, 45);//we position ourselves at the lower left
//corner. Nothing is drawn yet.
LineTo ( 10, 25);// a line is drawn up to the top left
LineTo ( 30, 40);// a line is drawn to the middle of the M
LineTo ( 50, 25);// a line is drawn up to the top right
LineTo ( 50, 45);// a line is drawn down to the bottom right

return 0;
}
```

We have just drawn the letter *M* by moving to a point on the grid and then drawing a sequence of lines from that first point through the other points we saw in Figure 8.5.

8.3 SOME HELPFUL HINTS WHEN DRAWING

In addition to some of the functions we just examined, we need to consider additional functions that help us draw. One of the understood features of drawing when using these functions is that you draw with an imaginary "pen." The "pen's" thickness and color can be controlled by the programmer through functions that we will examine next.

USING THE "PEN" TO DRAW

Before using the pen, we will want to set the size of the pen tip. Imagine that the tip of the pen is a rectangle rather than a point. How big that rectangular tip is will determine the thickness of the lines we can draw. Imagine the differences among pens whose tips are rectangles with these sizes:

2×2
4×6
8×10

The smallest tip will be the 2×2 tip and it will not be as big a tip as the 8×10 (Figure 8.6).

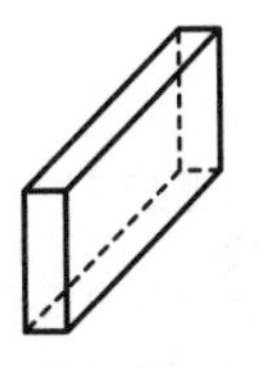

A line segment drawn with a "tip" that is 4 units wide and 6 units high.

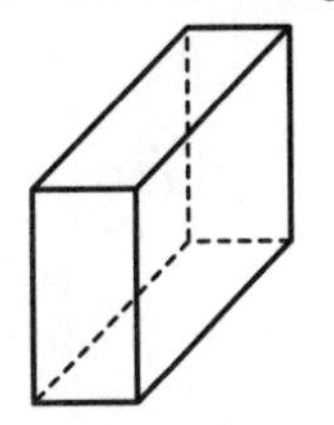

A line segment drawn with a "tip" that is 8 units wide and 10 units high.

FIGURE 8.6 Lines drawn with pen tips of varying sizes.

We set the size of the pen tip through a function called *PenSize*. This function takes two integers as parameters—one to set the height of the rectangle and one for the width. Here is the heading of the function followed by some calls to the function.

```
void PenSize ( int width, int height);

PenSize  (2,2);
```

```
PenSize  (4,6);
PenSize  (8,10);
```

Later in Section 8.5, we will look at a function that allows us to change the color with which we draw. For now, you may assume that we are drawing black lines on a white background.

BACKGROUND VS. FOREGROUND

It is important to define two terms in graphics and these are the background and foreground. Our background represents the window onto which we are drawing. Usually, it is white. The foreground represents anything we draw on the background. In Section 8.5, we will look at the functions that allow us to control the foreground color.

HINT!

It is important to familiarize yourself with the graphics functions that are available to you. These functions will be dependent on the hardware of your computer and its operating system.

8.4 HOW TO MAKE AN OBJECT MOVE

A nice thing to do after creating an image is to make the object *move*—that is, *appear in a new position*. Making an object move uses an interesting algorithm. We begin by displaying the object in some forecolor. (Remember that the *forecolor* is the color with which we draw against the background color—usually white.) Then we "erase" it by drawing it again in the same color as the background. This has the effect of making the object "disappear." After that, we draw the image again in its new position. It will appear to have moved!

Algorithm to make an object move

1. Display the image. (Use the code that creates the image.)
2. Change the color of the pen to the same color as the *background color* that you were drawing on.
3. Redraw the image in the background color.
   (The object will have seemed to disappear.)

4. Delay the computer.
5. Change the pen back to the original color.
6. Redraw the image using new points for positions.

Every time you move the object, you need to "erase" it and then redraw it in a new position.

SLOWING DOWN THE COMPUTER

Once you know how to move an image, you may need to control the speed of movement. Here it might be useful to employ an "empty for loop." This is a loop that spins nothing, but keeps the computer busy. Nowadays, most computers' microprocessors are so fast that the computer will be able to execute your program very quickly. In the case of moving an image, the computer will execute your instructions perhaps too quickly—so that any viewer will not recognize what you were trying to do. What we want to do is "delay" the computer. The computer can process and execute the directions faster than we can see the image.

DELAYING THE COMPUTER

To delay the computer as we mentioned previously, we need to keep the computer busy doing nothing—literally spinning its wheels. Think of all the times when you have had to wait for someone because you were ready and he was not. You usually try to occupy your time doing something like reading or watching people— because there is nothing else to do.

One of the easiest things to let the computer do to keep it busy is give it an "empty" *for loop* to execute. An "empty" *for loop* "spins" but does nothing. (Think of the analogy of lifting a bicycle wheel and then turning the pedal. You see the wheel spin, but the bike can't go anywhere because the wheel is not on the ground.) You write the code for the *for loop* but the loop itself is *empty*—that is, *the body of the loop has no programming statements*. The compiler then spends time setting the initial variable, checking the boolean condition using that variable and then increasing the value of the variable. It repeatedly does these three steps but because there is nothing to do inside the loop itself, the user does not see anything happen on the screen.

Let's look at the syntax of an empty *for loop*. We will set the control variable to *1* in each example but we will change the upper limit used in each boolean condition. The higher the upper limit, the more times the loop "spins."

EXAMPLES

```
for (int x = 1 ; x < 1000 ; x = x + 1) ;

for (int x = 1 ; x < 100000 ; x = x + 1);
```

In the first example, the loop spins 1,000 times. (That's a lot.) However, in the second example, the loop spins 100,000 times. (That's really a lot.) Notice in each example that a semicolon (;) follows immediately after the statement to show that the loop ends right away. This design shows that the loop is not controlling any other statements—it has no statements (called the body of the loop) to spin.

The best way to see the results of using an empty *for loop* is to run your program without this loop. If what you see appears to be happening too fast, then maybe you will need to slow down the program. This loop can be helpful for you.

8.5 USING FUNCTIONS TO DRAW

Most applications of a programming language will provide some graphics capabilities in the form of a toolbox or a library of functions. You need to access that library and look at those functions. The *frustrating* thing, however, is finding out what functions have been provided for your use. Depending on what version of what language you purchase, your graphics functions will be specific to that version and the *hardware* on which you run your programs.

You need to look at the functions themselves, and examine their headings—just like we did with *MoveTo* and *LineTo*. Once you understand what parameters (the variable types each function needs) and what each function does, you can make decisions about how you wish to draw some object.

In this section, we will examine some *generic* functions that are simulated in languages that provide graphics capabilities. You might be able to use these exact same functions or something like them.

PenSize (takes the width and height of the pen)
ForeColor (allows you to set the color of the foreground)
SetRect (this function sets the coordinates of each of the corners of a rectangle, but does not draw it)
FrameRect (this will draw the outline of the rectangle)
PaintRect (this will paint the rectangle with the forecolor you chose)

FrameOval (this will draw an outline of a circle/oval)
PaintOval (this will paint the circle/oval with the forecolor you chose)

First, we need to explain how each of these functions work. Then you can use them to make whatever drawings you wish. A rectangle has four sides. It is a special type of variable whose syntax we will explain later. Each side of the rectangle has a name that represents either a row or a column (Figure 8.7).

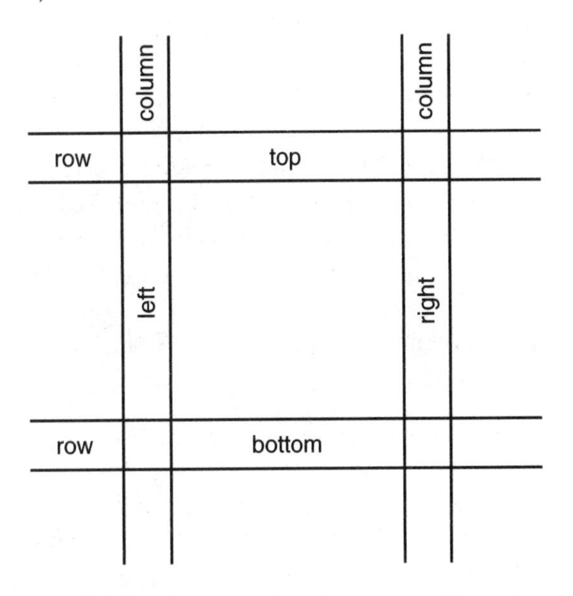

FIGURE 8.7 Each side has been named and drawn as either a row or a column.

Both the *top* and *bottom* of the rectangle are *rows* that you need to estimate depending on the size of the screen on which you are drawing. Both the *left* and *right* sides of the rectangle are *columns* that you position. There are two things to consider when you construct a rectangle: its position relative to the screen and its dimensions (width and height).

If a screen is 600 columns by 400 rows then you might want to position a rectangle near the middle of the screen. You would start to think about being at Column 300 and Row 200. If we want the rectangle to be 50 units wide and 100 units high, then we will start to consider the following positions. The right side of the rectangle should be at Column 350. If we want the rectangle to be 100 units high, then we should set the bottom of the rectangle to be 100 units below Row 200, at Row 300.

The difference between the two sides (left and right) of the rectangle is 50 units (350 – 300). The difference between the other two sides (top and bottom) is 100 units (300 – 200). Every time you construct a rectangle you need to consider where you will position it and then how to adjust the sides of the rectangle so that they have the appropriate dimensions that you want.

Once you decide which rows and columns you will use for your rectangle's sides, then you set their values by calling the *SetRect* function. You need to follow the heading given in *SetRect,* which might be like this:

```
void SetRect (rect * R; int Left, int Top, int Right, int Bottom);
```

Notice that the parameters of the function are all integers except for the first parameter, which is a rect. *But what is a rect?* It is a special variable (called a pointer variable), which we will examine later in Chapter 12. For now, just think of it as a "rectangle" and follow the syntax used in the example.

We want to make a call to *SetRect* but we don't have a name for our rectangle yet. So let's call it by this name ("house") and we will declare it in the following manner:

```
rect * house;// use the '*' in between the rect and its
// name
```

Next, we make the call by using the *integers* that comprise the sides of the rectangle:

```
left = 300;
top = 200;
right = 350;
bottom = 300;
```

Our call would be this:

```
SetRect (house, left, top, right, bottom);
```

Notice that, in the call, there are no types mentioned (i.e., *int* or *rect* *)—only the *names* of the variables are used. After calling this function the rectangle is ready to be drawn because all its coordinates have been set. Now, to draw it, we need to call another function, *FrameRect*. *FrameRect* has only one parameter, the rect itself, R. Let's look at its heading.

```
void FrameRect (rect * R);
```

Now we will *call FrameRect* in the main program using our rectangle called "house." Notice that, because the function is a function that does not return anything (note the word "void"), the call looks like this:

```
FrameRect (house);
```

The next thing we should do is fill in the house with some color so that it is not just an outline of a house. We will call the function *PaintRect* next. Here is its heading and the call we should use with it. Notice the differences in syntax between the heading and the call.

```
void  PaintRect  (rect * R); // function heading
```
return type   function name   parameter type   parameter name

```
PaintRect (house); // function call
```
function name   variable from main

WATCHOUT!

Calls always differ from headings because the *type* of variable is not mentioned in the call, only the *name* of the variable is mentioned. The compiler will "know" what type is being sent in because it has already translated the heading.

Now let's take care of some of the smaller details that we need to consider. Here is a list of typical colors from which you can choose. We will pick a color with which we will draw the house. Let's pick blue.

Cyan
Magenta
Green
Red
Yellow
Blue
Black
White

Let's look back at the heading that set the color for drawing. It was called *ForeColor*. It sets the color in the foreground—the area where we are drawing everything. For the second time, you will see an unfamiliar type. After the rect * type we saw in *SetRect,* we are now seeing the color type. Think of the color type as a built in type that is independent of the language. It really was defined for the sake of the computer's hardware. Because we need to pay attention to syntax, let's follow the syntax given in *ForeColor's* heading so that we can make an appropriate call:

```
void ForeColor   (color       c);// function heading
```
no return type   function name   new parameter type   parameter name
```
ForeColor (blue);  //function call
```
function name   color

One last detail is to set the size of the pen with which we are drawing. The pen tip size is measured in width and height. Let's set the width of the pen tip to be 4 units wide by 6 high.

```
void PenSize  (int width, int height);//function heading
```
no return type   function name parameter type   parameter name
```
PenSize (4, 6); // function call
```
function name   two integers

Notice that in the call to *PenSize,* no variables were used—only the direct values, *4* and *6.* We can do this when we are using value parameters. Recall that the function just needs the values of the variables and not the variable holders themselves. Now we will put together all the separate parts into one program fragment in the main function:

```
rect * house;
int top, bottom, left, right;

left = 300;
top = 200;
right = 350;
bottom = 300;
PenSize (4, 6);
ForeColor (blue);

SetRect (house, left, top, right, bottom);
FrameRect (house);
PaintRect (house);
```

FIGURE 8.8 A blue rectangle is shown representing a house on the CD-ROM.

ON THE CD-ROM

Look for figure 8.8 on the CD-ROM; that should draw a house that looks like the following.

So, in your algorithm for drawing any object, you need to set the color of the objects being drawn. You will probably have to change the color each time you draw something new.

DRAWING THE SUN BY USING THE OVAL COMMANDS

Next, if we wish to draw the sun in the sky, for example, we need to call a function that allows us to draw a circle. Circles are drawn through oval functions, which are functions that can draw any oval of any size specified. The size of the oval is determined from a circumscribed rectangle—that is, a rectangle drawn around the oval. First the rectangle's size is set according to the previous sides we mentioned (top, bottom, left, and right). Then, the largest possible oval (one that touches all sides) is drawn to fit inside of it (Figure 8.9).

Here are the headings of a couple of oval functions. By using the *SetRect* function that we saw previously, we can set the coordinates for the

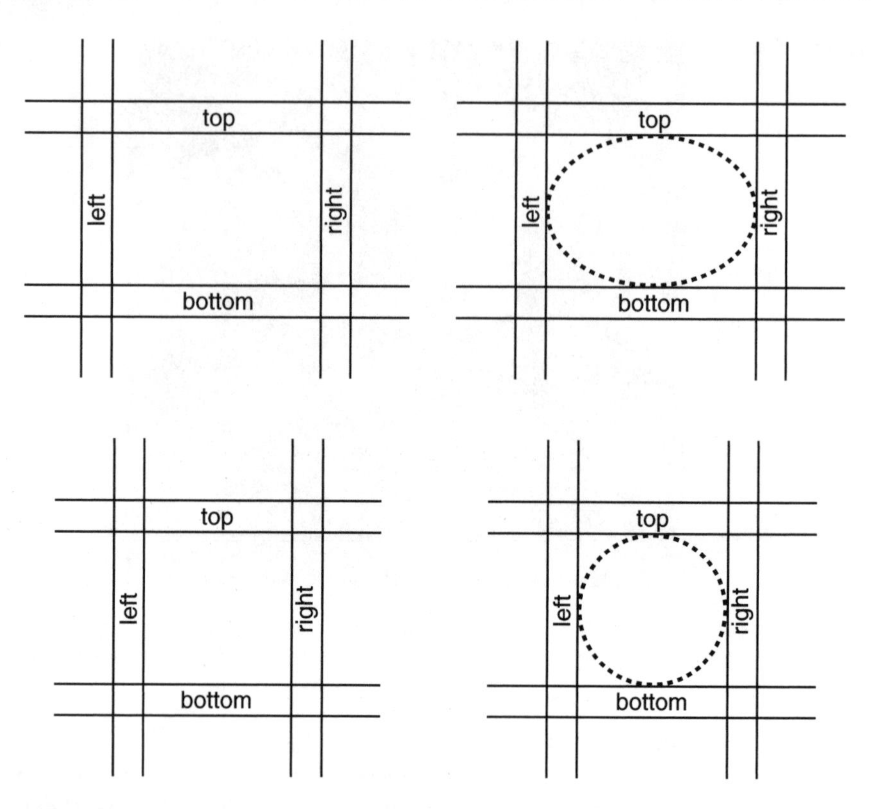

FIGURE 8.9 Two rectangles are shown empty and then with the largest oval that fits inside each. In the second example, the oval is a circle because the rectangle in which it is drawn is a square.

sides of the frame of the rectangle and then call on the oval commands to actually draw the oval. Here are the headings of the oval functions.

```
void FrameOval (rect * r);
void PaintOval (rect * r);
```

Before calling *SetRect,* we will set the coordinates of the rectangle in which we will draw the oval representing the sun. We will position it above the house and to the left of it by using the values that follow.

```
rect * sun;
int top, bottom, left, right;

left = 200;
top = 75;
```

```
right = 225;
bottom = 100;
//we will change the forecolor to yellow so that it looks
//appropriate.
ForeColor (yellow);
SetRect (sun, left, top, right, bottom);
FrameOval (sun);
PaintOval (sun);
```

Examining the rectangle's width, we can see that it is 25 units wide (225 – 200) and the length is also 25 units long (100 – 75). This demonstrates that the rectangle is really a square (25 units by 25 units). The oval drawn inside that square will be a circle.

Just from these two examples, you can see that it is a lot of work to call on these functions to draw objects. You also need to work within the parameters specified by the function. Imagine that drawing a circle involves designing a rectangle that is really a square so that an oval drawn inside it appears to be a circle. That is complicated! The good part about these functions is that they are good practice for working within a function's parameters.

EXERCISES

Assume that your screen is 200 columns by 100 rows. Complete each question so that a square (40 units by 40 units) is drawn exactly in the middle of the page.

What will be the value of each of these variables?

1. left
2. top
3. right
4. bottom
5. *Declare* a rectangle called *box* and *call* SetRect so that *box's* coordinates are established.
6. Call FrameRect and PaintRect .

SUMMARY

First, we began by defining graphics as programming the computer to draw shapes. We introduced a type of graphics called vector graphics where lines are drawn with direction in mind. A line is always drawn *from an initial position to a new position* where it ends, and for this reason is an example of vector (directed) graphics. Most graphics will be done by calling functions that do the work for you. The only thing to be careful about is what type of parameters each function needs to operate properly. Functions will vary from language to language and will also be dependent on the operating system used.

The first two functions we examined were *LineTo* and *MoveTo,* used to draw lines from your present position to a new position and for moving to a new position.

There are certain tricks that allow the programmer to make a drawing disappear.

The first point is to draw the shape and then redraw it in the background color. That has the effect of erasing it. The second point is to move the shape by drawing it in a new position. Every time you want to move, you have to draw the shape in the background color so that it can "disappear." If any of these tricks are executed too quickly by the computer, there is an additional tactic to be employed— "delaying" the computer. This is done by inserting an "empty" *for loop* to make the computer "spin its wheels."

There are many graphics functions that allow the programmer to draw different shapes on the screen. By calling these functions properly, the programmer only has to decide where a particular shape should be drawn. The screen is divided into rows and columns. Depending on the number of rows and columns visible on your screen, you will position shapes accordingly.

Rectangles are used in many of these functions and consist of a top and bottom, both of which are rows, and a left and right (both of which are columns). Functions such as *SetRect*, *FrameRect*, and *PaintRect* will set the size of the rectangle, then draw its outline and fill it in with the foreground color. The oval functions (*FrameOval* and *PaintOval*) are extensions of these rectangle functions by allowing the largest possible oval to be drawn within a rectangle whose position is set by the programmer.

ANSWERS TO ODD-NUMBERED EXERCISES

8.5

1. left 80
3. right 120
5. rect * box;

```
SetRect (box, left, top, right, bottom);
```

CHAPTER

9

Running Out of Holders? It's Time for the Array.

IN THIS CHAPTER

- The Array—Keeping What's Alike Together
- Members Bound by Their Type
- Loops Work Well with Arrays
- Programming with Arrays
- Assigning an Array
- Printing an Array
- Copying an Array

In this chapter, we will look at a data holder that is very useful in programming languages. It is called the array. The array is used to hold several values of the same type of variable. Let's start with some analogies before we look at the array itself.

9.1 AN ANALOGY FOR THE ARRAY HOLDER

The array is designed to hold a collection of data. When we deal with a large collection of data, we need to put each piece of data into a separate variable holder. If you have, let's say, 50 data items, you would have to sit and think of 50 variable names—one name for each data item. That can be difficult because there are only 26 letters in the alphabet. You could certainly use a, b, c, d, and so on, but what would you use after that? The array type holder provides a useful technique for holding large quantities of information. To better understand the concept of an array, let's look at an analogy.

Think of a large chest of drawers. Most bureaus have only three or four drawers. Imagine if you had a bureau with 50 drawers. You could refer to each drawer by a number—Drawer 1, Drawer 2, Drawer 3, and so on. In each drawer you would put clothes, but the clothes would differ from drawer to drawer.

Socks would be in Drawer 1. Shirts would be in Drawer 2. In Drawer 3, you would put sweaters for spring. In Drawer 4, you could put shorts. In Drawer 5, you could put jeans. Each drawer contains different items—socks, shirts, sweaters, shorts, and jeans—but each of these items is an item of clothing. All of these items belong in the bureau because they are clothing items.

An array is similar to a bureau—it has many different drawers (slots), called *members*. Each member is identified with a number. The array is used to hold *different values* that are of the *same type*. You could have an array of characters. Let's look at a bureau that has a different characters in each drawer. Compare it with the bureau of clothing we just discussed (Figure 9.1).

HINT!

In the programming language C++, the array "drawers" (members) are numbered beginning with 0 instead of with 1.

AN ARRAY IS USED FOR A COLLECTION

The array is used for a collection of items. Each item is unique, but does belong to a common *type*. Think about a collection of CDs. If we were

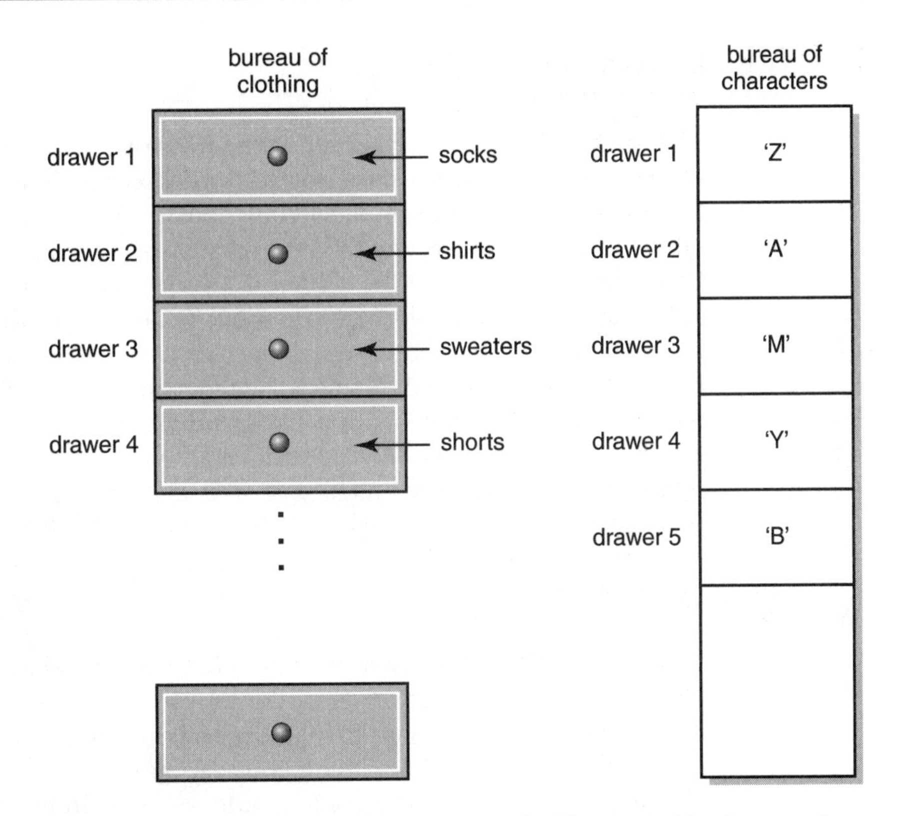

FIGURE 9.1 Two bureaus are shown—one with different clothing items and another with different characters in each drawer.

really organized, we might put a number on the case of each CD we own. Then we could generate a list of CDs such as the following:

CD 1: Henry Smith's *All-Time Hits*
CD 2: Pearl Jam's *Vitalogy*
CD 3: U2's *All That You Can't Leave Behind*
CD 4: *The Best of Beethoven*
CD 5: *Rocking to Bach*

The collection is a collection of CDs and each CD is unique. Rather than referring to each CD individually by its title, we refer to each CD by its number in the collection, CD1, CD2, CD3, and so on. What we have created is an array of CDs—a collection of CDs.

9.2 WHY IS THE ARRAY USED?

The array is used to group together variables of the *same type* but differing in values. As you learn to program, you may need to deal with large quantities of data. For example, if you know that you will have to store 50 names of people you know, then the array is a good choice of holder for this information.

The array would allow you to put each name into a drawer of a bureau, so to speak. Instead of calling each drawer, a drawer, we'll call it a name—Name 1, Name 2, Name 3, Name 4, and so on. Otherwise, you would have to come up with a list of unique variable names such as *first_friend, second_friend, third_friend,* and the like. When we get to array syntax, you will see how much easier it is to create 50 variables that are unique without having to type a variable name for each.

ARRAY SYNTAX

When you *declare* an array to the compiler, you are telling it to create a large bureau with each drawer of the bureau numbered. After you declare the array, you can start to load the array by assigning each member ("drawer") a value. The key thing about array syntax is letting the compiler know how many drawers it should set aside in memory and *what type item* it should expect *to store* in each drawer.

So far we have used the int, char, double, string, and boolean types. We could make an array of each of these types if we wanted.

EXAMPLE

Let's look at a declaration of an array of 50 members where each member is an integer. Just as we did with our collection of CDs, by naming each CD, CD1, CD2, and so on, we need to come up with a name for the collection. Let's use the term *Group.*

```
int        Group        [ 50 ] ;
```

type   array name   number of members

Notice that these brackets [ ] were used to let the compiler "know" it should create 50 integer holders. If the brackets had been missing, the declaration would have looked like a straightforward integer declaration.

```
int        Group ;
```

That would have meant that we were declaring one integer called *Group*. This array allows us to create 50 integers, all with the common name *Group*. Here's how these integers would be named. I need to forewarn you that, in the programming language C++, the array members are numbered *beginning with zero*. Let's look at how each variable is named—first in English and then in a programming language.

English	*Programming Language*
Group 0	Group[0]
Group 1	Group[1]
Group 2	Group[2]
Group 3	Group[3]

{Use the same format for Group 4 through Group 46.}

Group 47	Group[47]
Group 48	Group[48]
Group 49	Group[49]

There is no Group 50 because Group 49 was the 50th member of the array. (That's because we started at Group 0 instead of at Group 1.) The brackets are used to identify the member number ("drawer number") of the array. If the brackets were not used, the compiler would "think" it was dealing with an integer. We are emphasizing that the collection is called "Group" and each variable in the collection is referred by its number in brackets ([ ]) next to the name. Let's put a different integer into each member (Figure 9.2).

HINT!

Note that values have no correlation with the number member where they are stored.

DIFFERENT TYPES OF ARRAYS

Let's declare some other arrays using different variable types. We'll start with an array of strings to hold a collection of phone numbers. We'll declare an array of 35 strings.

```
string   Phone_Book       [35] ;
type     array name       number of members
```

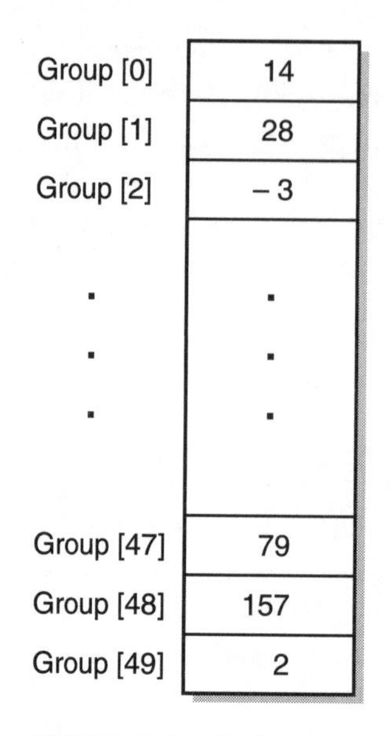

FIGURE 9.2 Each member of the array is shown with its value inside of it.

Another example of an array would be one used to hold a collection of real numbers. Imagine that you were trying to store the checking account balance for a group of 40 individuals. Each person would need a slot or drawer in the array for her balance.

```
double   Account_Balance      [ 40];
```
type   array name   number of members

ASSIGNING VALUES TO AN ARRAY

Now that we have declared a couple of different arrays, it is time *to load* an array. What that means is that we will *assign each member of an array* a value. Let's start with the phone_book we declared previously. Consider this list of members and the value each slot contains.

Member	*Should Contain*	*Value*
Phone_Book [0]		"641-2222"
Phone_Book [1]		"123-4567"
Phone_Book [2]		"654-2345"
Phone_Book [3]		"234-4567"

Phone_Book [4]	"890-1234"

.

{Use similar format for Phone_Book 5 through Phone_Book 30.}

.

Phone_Book [31]	"345-5678"
Phone_Book [32]	"567-6789"
Phone_Book [33]	"345-9876"
Phone_Book [34]	"432-4567"

We have only used 18 slots in the array; the other 17 are empty and have nothing in them—not even blanks. The compiler considers them to be *uninitialized*. This means that they have not been given any value. Remember that if you try to work with a variable that has no value, the compiler will have problems.

Now another point to notice about the array values in Phone_Book is that integer values have been mixed with the dash, ' - .' Yet, we know that this array has been declared as an array of strings. A string type is useful because you can mix characters with numbers and still store them in one variable holder.

The next point to consider is how we will assign values to members of an array. Of course, we could use several assignment statements to load (assign) the array:

```
Phone_Book [0] = "641-2222" ;
Phone_Book [1] = "123-4567" ;
Phone_Book [2] = "654-2345";
Phone_Book [3] = "234-4567";
Phone_Book [4] = "890-1234";
```

{Use similar format for Phone_Books 5 through Phone_Books 30.}

```
Phone_Book [31] = "345-5678" ;
Phone_Book [32] = "567-6789";
Phone_Book [33] = "345-9876";
Phone_Book [34] = "432-4567";
```

But, that seems to be a tedious way of assigning values to several variables. You might also start to think "what is the point of the array?" because we don't appear to be saving any time. The answer to this tedium is found in the *application of a loop* to our problem; a loop is ideally suited for *accessing members* of the array *efficiently*. The real power of the array is using it in conjunction with a loop to save time.

EXERCISES

Declare an array of each item with the given number of members.

1. An array called "Results" of 30 integers
2. An array called "Initials" of 14 characters
3. An array called "Amounts" of 100 real numbers

9.3 HOW TO USE LOOPS WITH ARRAYS

To understand how a loop can be used in conjunction with an array, let's start by comparing the members of the array with the control variable of a *for loop* (If you recall, the control variable is the variable that is increased or decreased during the loop). We'll start by declaring an array of 10 members, each of which is an integer. Next to the array, we'll declare a control variable that will be used in a *for loop.*

```
int   list [10];
int   x;
```

The first step in understanding how a control variable might be useful with an array is to think of the control variable as a variable that *moves through the members ("slots" or "drawers") of an array.* The control variable changes *during a for loop.* If we use it to move through the array members—one at a time—we can use a loop to access the members more efficiently. Consider this list of array members, slot position, and the control variable's values.

Member	Slot #	Control Variable Value
list [0]	0	0
list [1]	1	1
list [2]	2	2
list [3]	3	3
list [4]	4	4
list [5]	5	5
list [6]	6	6
list [7]	7	7
list [8]	8	8
list [9]	9	9

DESIGNING A FOR LOOP WITH AN ARRAY IN MIND

A *for loop* can be set so that its control variable spins *through a range* of numbers that corresponds with the slots of the members of an array. In the array we just declared, we have 10 members that are numbered 0 through 9. Now we need to set up a *for loop* whose control variable will spin from 0 through 9, inclusive. (The term *inclusive* just means that we will hit the numbers 0 and 9 in addition to the numbers in between.) Our *for loop* would look like this:

```
for (int x = 0; x <= 9; x = x + 1)
```

This *for loop* will spin 10 times and hit every number from 0, 1, ... , 7, 8, and 9. The next part of our problem is to address how to use the *for loop* to access members of the array. Consider each array member's name:

list [number]

The *number* is the only part that changes as you move through successive members in the *list*. Each array member has the name "list" followed by the brackets with the slot number inside. We'll use the control variable to represent that number. That way, each time the *for loop* "spins," a different member of the array is accessed.

If the variable x *has the value*	*then list[x] refers to*
⇓	⇓
0	list[0]
If the variable x *has the value*	*then list[x] refers to*
⇓	⇓
1	list[1]
If the variable x *has the value*	*then list[x] refers to*
⇓	⇓
2	list[2]
If the variable x *has the value*	*then list[x] refers to*
⇓	⇓
3	list[3]
.	.
.	.
.	.

If the variable x has the value	*then list[x] refers to*
⇓	⇓
8	list[8]
If the variable x has the value	*then list[x] refers to*
⇓	⇓
9	list[9]

Let's start with an example where we use the *for loop* to print out all the members of the array *list*.

```
for (int x = 0; x <= 9; x = x + 1)
{
cout << list[x]<< endl;
}
```

HINT!

Recall that the statement, x = x + 1, will cause the value in x to increase by 1. This statement allows the *for loop* to "spin." In this case, the counter statement, x = x + 1, allows us to look at each slot of the array during each "spin" of the *for loop.*

First, *list*[0] is printed, then *list*[1], *list*[2], and so on. Let's consider some other examples using the *for loop* to move through the array. In the next example, we will let the user assign the members of the array. We'll put a message to the user inside the loop, as well as an input statement. It will look like this:

```
for (int x = 0; x <= 9; x = x + 1)
{
cout << "Please type an integer."
cin >>  list[x];
cout << endl;
}
```

To understand everything that happens in the loop, let's consider the dialog between the screen and the user:

What the User Sees on the Screen	*What the User Types at the Keyboard*
`Please type an integer.`	5
`Please type an integer.`	18

```
Please type an integer.              22
Please type an integer.              -1
Please type an integer.              98
Please type an integer.              12
Please type an integer.              100
Please type an integer.              24
Please type an integer.              16
Please type an integer.              31
```

As a result of the *for loop*'s execution, the array looks like a chest of drawers with each of these values in the appropriate box (Figure 9.3).

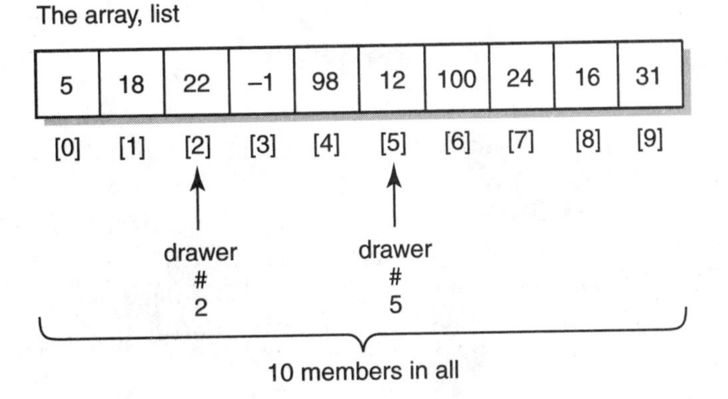

FIGURE 9.3 The array *list* is shown with its values—each in the appropriate drawer of the array. There are 10 members in all.

HINT!

Notice that the values in the array are not in any order. They were typed by the user. The array will not arrange these numbers in any order.

WATCHOUT!

There is no correlation between the array slot and the value in the slot. It would be unusual to have the value 10, in list[10] for example. That would be a coincidence.

For loops are the easiest way of moving through an array because you can set the control variable's values to match those used for the array's members. Here are some other examples of using *for loops* with the array.

9.4 SOME SHORT EXAMPLES USING THE ARRAY

It is important to look at the different examples that you will encounter when working with an array. You will want to be able to load (assign) an array, print its values, and initialize an array, for example. In this section, we consider some fragments of code that will accomplish these tasks and others.

INITIALIZING AN ARRAY

If you recall, when you *initialize* a variable, you give it its *first* value. When you initialize an array, you put a value (usually 0) into each slot of the array. For that reason, the drawers of the array are never empty and the compiler does not experience any problems when it accesses these members.
In the first example, we declare an array of integers and set all the members to 0. (Usually, when you initialize an array of numbers, you assign each member 0.)

```
int ages[15]; // declaring an array of integers

for (int count = 0; count <= 14; count = count + 1)
{
ages[count] = 0;//zero is being assigned to each slot
}
```

In the second example let's declare an array of characters and initialize each member to the blank character—the space. Here's a fragment to do this.

```
char name[20]; // declaring an array of characters

for (int x = 0; x <= 19; x = x + 1)
{
name[x] = ' ';//the blank is being assigned to each slot
}
```

PRINTING AN ARRAY

When using the *for loop* to print the members of an array that has already been assigned its values, decide how you want your output to look. If you wish the array members to be listed horizontally, look at Example 1. Example 2 shows the array with its values listed on separate lines. Note that in each version, we put the value in "x" to indicate which member of the array we are printing.

EXAMPLE 1

```
char friend [20]; // declaring an array of chars

for (int x = 0; x <= 19; x = x + 1)
{
cout << x << " " << friend[x]<< "  ";//info will be
// put on one line
}
```

EXAMPLE 2

```
char friend [20]; // declaring an array of chars

for (int x = 0; x <= 19; x = x + 1)
{
cout << x << " " << friend[x]<<endl;//info will be
// put on separate lines.
}
```

ON THE CD-ROM

A program which initializes, prints, and then lets the user assign the members of the array.

COUNTING THE NUMBER OF NEGATIVE VALUES IN AN ARRAY

Imagine an array of 100 real numbers where each real number stands for the balance in a checking account for one individual. If you work at the bank, you would probably want to keep track of the number of people who have overdrawn their accounts. In this fragment, we will count the number of people (array members that are negative) who have negative balances. If an array of doubles *has already been assigned* its values, we can use an *if...statement inside a for loop* to accomplish our task.

```
double balances[100]; // declaring the array
int total_count = 0;// intializing a variable
// to keep track of the negative numbers
                 . . .
for (int count = 0; count <= 99; count = count + 1)
{
if (balances[count] < 0)
   total_count = total_count + 1;
//total_count is increased each time a negative value in a
//member is recorded.
}
```

Imagine that the array already has its values, and that a small arrow is running down the right side of the array. The arrow will count the number of negatives it finds in the array. This total will be stored in the variable, *total_count* (Figure 9.4A and B).

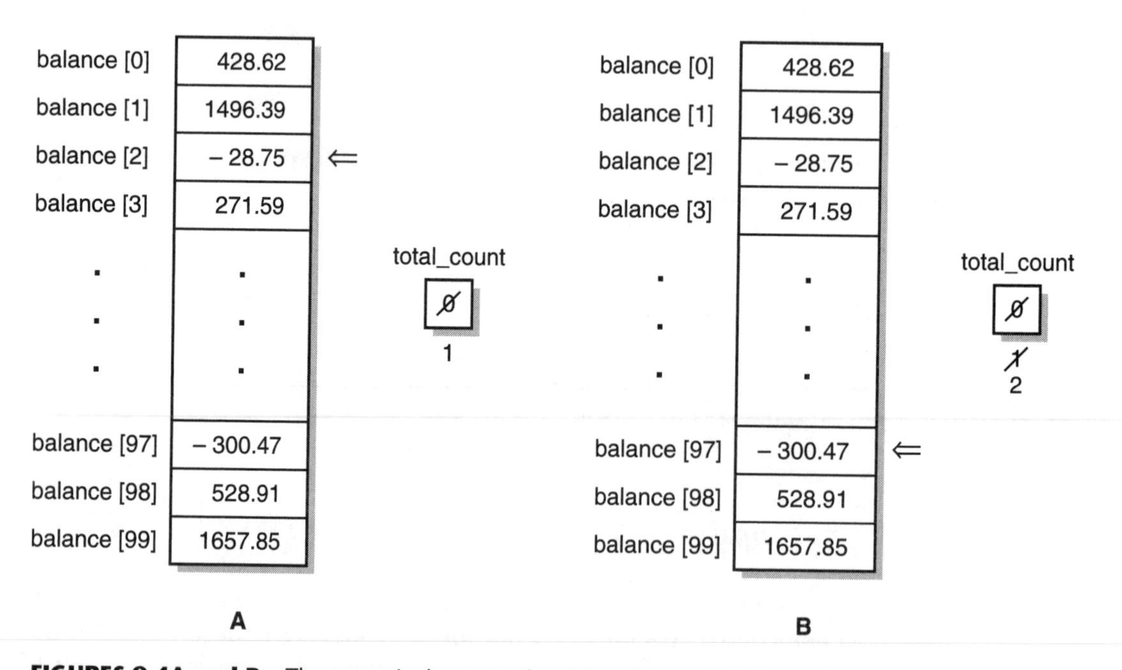

FIGURES 9.4A and B The arrow is shown to the right of the only two negative numbers it encounters. As a result, *total_count* has the value *2.*

COPYING ONE ARRAY TO ANOTHER

Let's say you are given an array that you wish to copy. First you need to declare another array (called *copy_array*) of the *same type and size*. Then you can use a *for loop* to copy each member from the original array into the "copy" array, one member at a time (Figure 9.5).

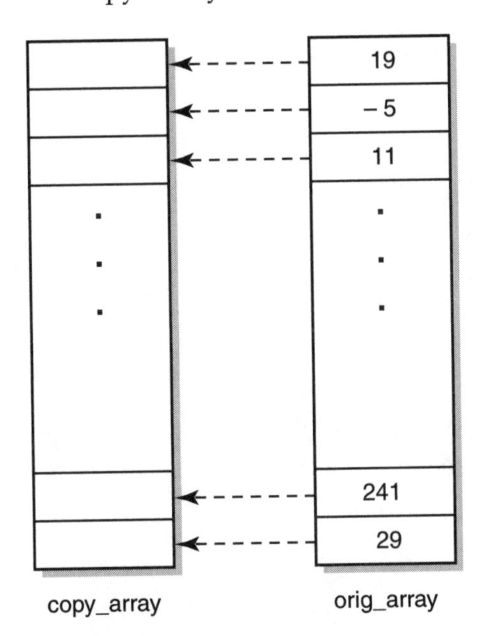

FIGURE 9.5 "orig_array" is shown passing copies of its values into the empty array, "copy_array."

Now consider the code that will execute this problem. Keep in mind that we are assuming that orig_array has already been assigned its values.

```
int orig_array [50];
int copy_array [50]; // declaring two arrays of ints
.
. // code to assign orig_array goes here
.
for (int y = 0; y <= 49; y = y + 1)
{
copy_array[y] = orig_array[y];
/* each member of orig_array is copied, one at a time, into the
  new array. This section presumes that orig_array was assigned
  already. */
}
```

ON THE CD-ROM

A program to swap all the members of one array with another.

EXERCISES

Declare an array of the appropriate size and type as stated.

1. A 12 member array of booleans
2. A 25 member array of strings
3. A 5000 member array of integers

Write a fragment of code in C++ (using a *for loop*) to do the following:

4. Let the user assign the array in Problem 2.
5. Assign the value 0 to each member of the array in Problem 3.
6. Assign the odd slot members of the array (1, 3, 5, 7, etc.) the value *true* and the even slot members (0, 2, 4, 6, etc.) the value *false* for the array in Problem 1.

SUMMARY

In this chapter, we introduced the array as a collection of holders for the same data type. An array could be a collection of integers, reals, booleans, strings or characters. An array is like a collection of drawers where each drawer contains a separate value, but all values are of the *same type*. We explained the array through the analogy of a bureau where each drawer of the bureau contains different kinds of clothing—socks, sweaters, shirts, and the like—but all are of the same type— articles of clothing.

When declaring an array, state the type first, then the array name followed by brackets, [ ], with the number of members indicated. The array declaration differs only slightly from other declarations by using these brackets. All arrays have members (like slots or drawers) that are numbered beginning with 0, then followed by 1, 2, 3, and so on. Take an array called *Group*, for example. The members are called *Group*[0], *Group*[1], *Group*[2], and so on.

Loops, especially the *for loop*, work well with arrays. Whenever you want to work with an array, such as assigning it values or printing its values on the screen, using a *for loop* to access the members of the array is quite useful. The *for loop* should be declared so that its control variable spins through the array members. For this reason, most *for loops* that are used in conjunction with arrays start spinning at 0 because arrays are generally numbered 0, 1, 2, and so on.

In the last section of the chapter, we covered different examples using the *for loop* with an array. We printed, initialized, and then copied one array into another array.

ANSWERS TO ODD-NUMBERED EXERCISES

9.2

1. int Results [30];
3. double Amounts [100]

9.4

1.
```
boolean  group[12] ;
```
3.
```
int collection[5000] ;
```
5.
```
for ( int m = 0; m <= 4999 ; m = m + 1)
collection[m] = 0 ;
// this for loop does not need braces since it has
//only one line of code to execute.
```

CHAPTER

10

But What if Things are Different?—Structures/Records and Fields.

IN THIS CHAPTER

- The Limitation of the Array
- What to Do If You Must Mix Data Types
- Definition of the Record and Its Fields
- The Struct—Short for Structure
- Programming with Structures and Records
- One Step Further—Arrays of Records/Structures
- Scope; Global and Local Variables

After having studied and worked with the array, you will notice that something is lacking in our exploration of data structures. That is, we do not have any type that can hold *mixed* types (i.e., something that could hold both an integer and a string at the same time). That is why we introduce the *record* in this chapter.

10.1 BEYOND THE ARRAY: THE RECORD

Once you have seen the standard data types—the integer, real, character, string, and boolean—you are ready to design your own data types. This includes the array whereby you can declare a *collection* of *one type*. Now we can introduce the *record* that allows you to create a *collection* of *different types*.

The *record* is used to hold various data. If you want to keep track of information about an individual, such as her name, age, and G.P.A., a record would be a good choice to hold this data. The name would be a string, the age would be an integer, and the G.P.A. would be a real number (because of the decimal).

Think of the record as a box into which you throw different things. Unlike the bureau analogy (used for the array in Chapter 9), where you only put clothing items into the bureau, the box—an analogy we use for the record—can hold anything that we define for it.

HOW TO DESIGN A RECORD

The first thing to consider about a *record* is what types of data you would like to put inside of it. In our previous example, we grouped together a string, an integer, and a real number for a record that keeps information about an individual (probably a student). Let's give names to each of these types:

```
string  name;
int   age;
double g_p_a;  //recall that variable names do not have any
//periods in them. You will soon see why.
```

This declaration shows each of these variables declared separately. The purpose of the record, however, is to group these variables together into *one new variable* of the record type.

DEFINING A RECORD

Before you define a record, you need to think of a good name for the record's contents. Let's name our record *individual* for the information we just mentioned. The syntax for declaring a record will include braces, { }, to show where the record *begins* and *ends*. Here is an example of a declaration in the programming language C++. A record in C++ is called a "struct" (short for "structure)," which suggests that you are building something unique.

```
struct individual
{
// the different data types go in here

}; // the semicolon ends the declaration
```

This is just a skeleton of a definition. We have not finished it yet. Inside the braces, we will *declare* the internal variables of the record—each of which is called a *field* of the struct (record).

```
struct individual
{
string name;
int age;
double g_p_a;
}; // the semicolon ends the declaration
```

Individual is the name of a box with three things inside of it—an individual's name, age, and G.P.A. (Figure 10.1).

RESERVED WORDS

Reserved words in any language (sometimes called "keywords") are words set aside because they have special meanings in the language. When the compiler (or interpreter) encounters a "reserved word," it will translate it according to the meaning already associated with it. Let's consider this brief list of reserved words that would be common to most programming languages: "for," "do," "while," "and," and "or."

Because these words have been reserved, a programmer will avoid using them as a choice for a variable name. Now that we have the skills to create our own variable types as well as names, it is important that we understand that we should <u>avoid</u> using these words. The compiler, when

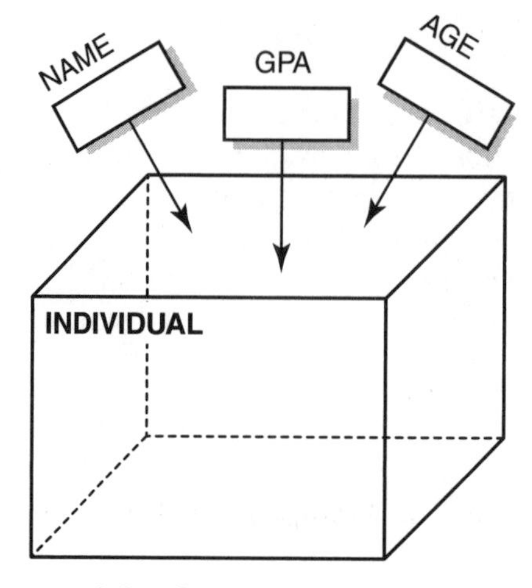

A Box Symbolizing a Record

FIGURE 10.1 A box is shown with the name "individual" on the outside. Inside it are the name, age, and G.P.A. *fields.*

it encounters a reserved word, will start to execute the command or commands associated with that word. If the compiler sees the word "for," it will assume that there is a loop that must be executed. This is just one example and it is the main reason why we *avoid choosing a reserved word as either a variable type or variable name.*

DECLARING ONE OR MORE RECORDS

Once you have *defined* the record, you are ready to *declare* one or more records. When you declare a record, you will be telling the compiler that a variable of type *individual* now exists. Remember—because of the declaration you made to the compiler—a new type, *individual,* now exists for the duration of that program (this is in addition to the standard types, int, char, double, etc.). Let's declare two variables of type *individual.*

```
individual first_person, second_person;

//Individual is not boldfaced because we want to emphasize
//that it is not a reserved word.
```

Once you have declared these two new variables, the compiler will set aside memory for each variable, *which includes* memory for each of its fields. The name you give that *individual* is the variable name.

DISTINGUISHING AMONG THE RECORD'S TYPE, VARIABLE, AND FIELDS

The most difficult part of dealing with records (called structs in C++) is the confusion that arises from all the variable names that are part of the definition. Let's take a moment to clarify, through drawings, to what each variable name, struct (record) name, and type refer.

When a record is *defined*, the compiler is informed what it will contain and, as a result, *sets aside* the appropriate *memory* for everything that the record contains. Each internal variable (each *field*) has its *own type* and *memory requirements*. Each record has to be given a name to identify it as a new "type." That way, a later declaration of the type is possible. Let's examine two different type records with similarities.

A *struct* that contains two strings and one character will be used to define a person's first name, middle initial, and last name. That *struct* will be given the name *identity*. *Identity* represents a new type. (C++ is a case-sensitive language so *identity* must begin with a small letter.)

```
" struct " indicates a new type        type name  // like int, char, etc.
        ⇓                                  ⇓
    struct                         identity//struct definition begins
{

string first_name;
   ⇑            ⇑
 field type     field variable name
string last_name;
   ⇑            ⇑
 field type     field variable name
  char  mid_init;
   ⇑            ⇑
 field type     field variable name
};
```

Another *struct* will be defined with an integer and two strings—the integer for the apartment number and the strings for the street address and city. (We assume all these locations are for people who live in apartments.) This *struct* represents a new type called *location*. For the duration of the program, we have seven types available—an int, string, double, char, boolean, *identity*, or *location*.

```
struct location
{
 int apt_no;
string street;
string city;
};
```

Defining a *struct* in the programming language C++ allows us to create a new type. This *struct* is really just a collection of variables—usually grouped together for clarity in the design of a program. In the *identity* type, two strings and a char are used to represent the full name of a person. The *location* type has an integer and two strings to represent a complete address. When you declare variables of these types, you declare them in the same manner that you declare any variables—you first state the type, followed by the variable names. Let's declare some variables of type *identity* and *location*.

```
identity  first_individ, second_individ;
location address1, address2;
```

Four variables have just been declared and all of them are, as yet, unassigned. These variables would be declared alongside other variables used in the program as in this example:

```
int x;
char initial;
identity  first_individ, second_individ;
location address1, address2;
```

EXERCISES (10.1.i)

Define a *struct* with the suggested fields in each problem.

1. A "movie" type that contains fields for the name of the movie, the year it was released, and the money it grossed (in millions) when it was released.
2. A "baseball_ player" type that contains fields for the player's name, his batting average, his homeruns, and his R.B.I.'s.
3. A "contact_info" type that contains fields to represent a person's home phone number, cell phone number, and e-mail address.

HOW TO ASSIGN RECORD VARIABLES

When you assign any record variable, you need to assign each field of the record. The record is, essentially, a group of other variables—the fields. When you assign a record or struct, you are really assigning its fields. It is important to both recognize the *field name* and distinguish it from the *field type*. To assign a record, we focus on the field name in the assignment statement. Consider this example where we *partially* assign three fields from an "*identity*" type.

```
first_name = "Mike";
last_name = "Smith";
mid_init = 'T';
```

To complete the assignment statement, we need to use the *variable name* of the "*identity*" type. In that way, we will be communicating to the compiler that we understand the connection between the field name and the new type. First, we need to examine the meaning of a *delimiter*, which will allow us to make this connection for the compiler. Using a *delimiter*, we will access the field names through the the variable name.

USING A DELIMITER

A *delimiter* is a punctuation symbol used to get access to an internal part of a larger part. In the case of a record or struct, the record or struct is the larger part, and the field is the smaller part. The delimiter is used to connect the variable name to the field name. When you want to access the fields, you use the variable name that was declared, followed by a delimiter, and then the field name. You can use this sequence in an assignment statement or any other acceptable statement for the variable.

Let's look at an example using the variable name, *first_individual*, and one of its fields—*first_name*. We will use the delimiter (a period—' . ') between the variable name (the name used for the new type) and the field name.

```
first_individ.first_name
   ⇓                 ⇓
the variable name    the field name
```

Notice that we have not used the "type" names in this example. Type names are only used for declarations and definitions. Consider the next example, which is a complete assignment statement.

```
first_individ.first_name = "Mike";
   ⇓               ⇓           ⇓
```

the variable name the field name the value assigned to the field "first_name" found in the variable "first _individual"

In this example, we are *really assigning the field* of the variable. When all the fields of the record have been assigned, then the record is considered assigned or "loaded." Let's assign the other fields of this struct.

```
first_individ.last_name = "Smith";
first_individ.mid_init = 'T';
```

HINT!

Periods are often used to separate larger parts from smaller parts. Just consider this fictitious World Wide Web address www.mycompany.com. Notice how the period is used to separate the company name "mycompany" from the "com" or the entire commercial group of companies on the Web.

LETTING THE USER ASSIGN A RECORD

The user can assign a struct by responding to the "cin" statement. We will follow the syntax used in the previous examples. Instead of using an assignment statement with the operator, "=," we will use the "cin" statement. The user should be given some prompt so that she understands what input she should be typing. Here are some examples using the variable name, *"address_1"* followed by all the field names in the *"location"* type.

```
cout << "Please type the street address."<< endl;
cin >> address1.street ;
cout << "Please type the city." << endl;
cin >> address1.city;
cout<< "Please type the apartment number."<< endl;
cin>> address1.apt_no;
```

Now the record *address1* has been *completely* assigned because *all* its fields were assigned. (Remember that when you declared *"address_1,"* it acquired all the field names and types associated with the new *location* type.

The same thing can be done for *address2*—either the user or the programmer can assign it. Right now, only *first_individ* and *address1* have been assigned; the other two record variables, *second_individ* and *address2*, are unassigned (Figure 10.2).

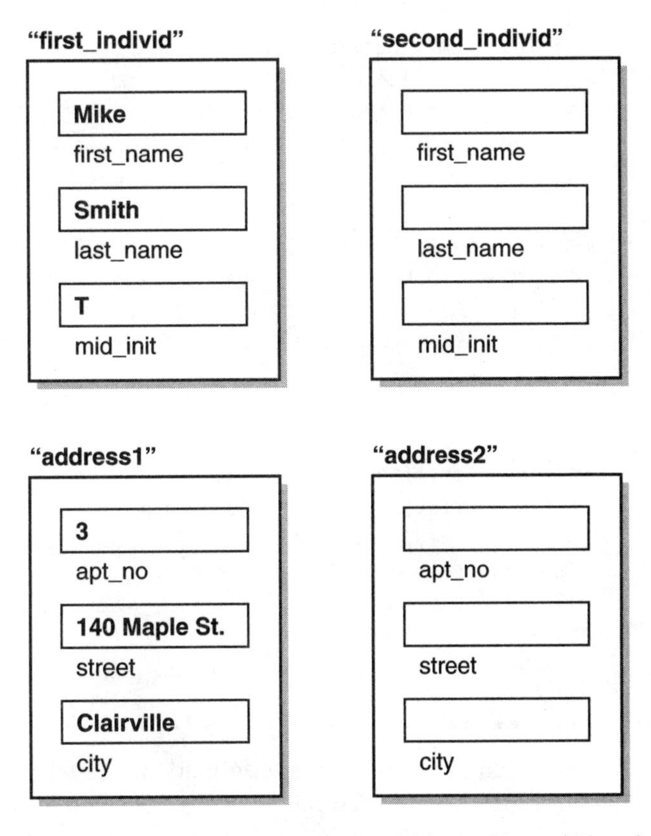

FIGURE 10.2 A diagram of the four record variables is shown. Either the fields have been assigned or they have been left blank to show they have no values yet.

EXERCISES (10.1.ii)

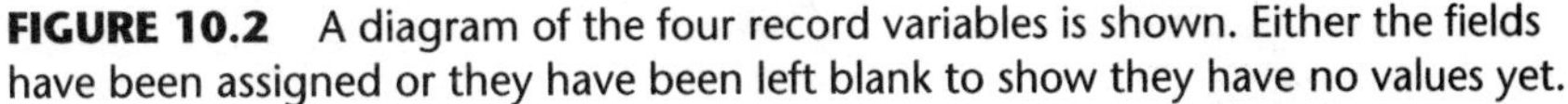

Assign each variable in each problem in the indicated manner.

1. *second_individ* by the user
2. *address_2* by the programmer

EXAMPLES OF DIFFERENT RECORDS

Records are useful for storing information about items with varied data. Think of the inventory in a store for each item sold. A piece of clothing, for example, could have the following fields to describe it:

```
struct item
{
string item_type;//pants? shorts? shoes? etc.
int floor;//floor where item is found
double price;//the price of the item
int percentage;//highest percentage discount allowed on //item
};
```

An item could be a pair of jeans, found on Level 2 of the store and priced at $48, but with an allowable discount of up to, but no higher than, 50% off. Another example is a struct that contains information about a CD in a music store.

```
struct CD
{
string name;//the name of the CD
string artist;//the artist's name
double price;//price of CD
int year_released;//year CD came out
};
```

EXERCISES (10.1.iii)

For each of the previously defined structs, declare a variable of that type and then assign each field of your declared variable.

1. item
2. CD

Declare a variable of that type and let the user assign each field of your declared variable.

3. CD
4. item

Identify the name of the new type that has been created.
Then name each field and its type in each struct.

5.

```
struct student
{
string name;
int grade_level;
int year_of_grad;
};
```

6.

```
struct movie
{
string name;
int year_released;
double revenue;
};
```

7.

```
struct book
{
string title;
string author;
int year;
char discount;
};
```

10.2 AN ARRAY OF RECORDS

Usually, when we define a record, we need it because we are anticipating a program that requires variables for lots of data. Think about the problem of storing all the pertinent information in a collection of CD's, books, or videos. You are not really designing a record data structure because you have only one CD in mind. You are really contemplating storing the data for *all* of your CD's. What we need to define is an *array* of records where each member of the array is a record (Figure 10.3).

DEFINING AN ARRAY OF RECORDS

We define an array of records just as we would define any array. But first, let's review a definition that we saw previously—the definition of a CD. Let's start by restating that definition, which was a struct.

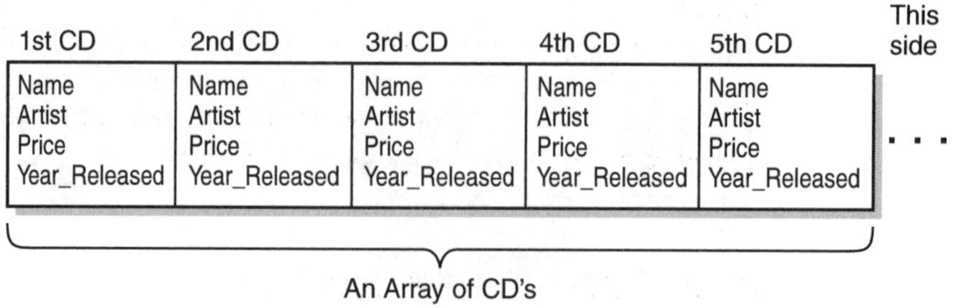

1st CD	2nd CD	3rd CD	4th CD	5th CD	This side
Name Artist Price Year_Released	Name Artist Price Year_Released	Name Artist Price Year_Released	Name Artist Price Year_Released	Name Artist Price Year_Released	. . .

An Array of CD's

FIGURE 10.3 An array of records is shown where each slot of the array is a record.

```
struct CD
{
string name;//the name of the CD
string artist;//the artist's name
double price;//price of CD
int year_released;//year CD came out
};
```

Next we will declare an array of CD's. The first step is to state the type (CD), followed by the name we wish to give the array, and followed by brackets with the number of array members inside. Remember that the array declaration is like any other declaration except that the brackets are used with the *number* of intended members inside.

```
CD my_collection[25];
```

⇑ ⇑ ⇑
type array name number of members

Each member of the group is a CD struct. Look at this list of variable names on the left and types on the right.

Variable Name	*Type*
my_collection[0]	CD
my_collection[1]	CD
my_collection[2]	CD
my_collection[3]	CD

HINT!

Always define a record (struct) *before* declaring an array of that type. Otherwise the compiler will not understand what the array contains.

WHERE TO PLACE THE RECORD DEFINITION

You need to precede the array declaration by the record (struct) definition. Otherwise the compiler will not "understand" what you are declaring. Because our example has been written in the programming language C++, we will refer to our record as a "struct." Recall that you are creating a new type when you define a struct.

If you think that you will be writing functions that use the struct definition, you will need to put the definition in a place where everything can "recognize" it. For this reason, we place the struct definition at the top of the program.

Consider this skeleton of a program. Notice that we have placed the struct definition at the top of the program—above the function headings and the main function.

```
#include <iostream.h>
#include <string.h>

struct book
{
string title;
string author;
int year;
char discount;
};

void Print_ALL ( book B);
void Change_Discount (book & B);

int main ()
{
// declaring three variables of the book type
book  my_book, your_book, x ;
.
.
.
```

```
return 0;
}
```

The *book's* definition is placed above every other part of the program where a variable could be declared. The reason for placing it here, as opposed to placing it just before the declaration, is to allow all functions, including the main function, to recognize this new variable we have just created. Our next topic concerns the *scope* of a variable; this topic will help you to understand this point.

Every part of the program should recognize the struct definition of a book. For this reason, we place the struct definition above all functions. By placing the definition here, we are making the struct definition *global*, that is, the definition is *recognized globally* or *everywhere within the program*.

10.3 GLOBAL VARIABLES, LOCAL VARIABLES, AND THEIR SCOPE

There are other ways to classify variables than by what type they are. Variables can be classified according to the *extent* of their *recognition*. If the variable is recognized by every function including the main function, then we say it is a *global* variable. Think of some of the most popular actors, rock stars, and politicians. Some are recognized only locally whereas others are recognized everywhere or *globally*.

Think of the example of a U.S. Senator. She would not be widely recognized outside of her own state. Now take the Majority Leader of the Senate. Many people would recognize him because his popularity would extend further than the state that elected him. However, he would probably not be recognized outside of the United States. Next, consider the example of the President of the United States. Most people would recognize the name of the President of the United States.

Each of these individuals has a *scope* of recognition. The term *scope* refers to the largest possible place where that individual is known. We use the term *local* to describe a limited scope—one that is not global.

Individual	*Recognized Where*	*Scope*
A U.S. Senator	the state that elected him/her	local to the state
The leader of the Senate	most/all states	local to the US
The President	most/all countries	global

Another example to consider is the scope of recognition for three soccer players. The first player is the best player in your town soccer league. Let's call him Joseph Thomas. Most people in the town recognize Joe because they have been watching him play since he was a child and they know how good he is. He is recognized throughout the town but *relatively unknown* beyond there. Now take a well-known college player. He plays on a college team and is so skilled that most college players have heard of him. His scope would be the college teams, including the players and those who watch their games. Now take Mia Hamm. She has played in the Olympics, the World Championships, and is now a professional soccer player for the first ever Women's Soccer League. Many people in many countries have seen her play and have heard of her. Her scope is much larger than the college player or the town player. We would consider her scope to be global.

Individual	*Recognized Where*	*Scope*
Town player	the town	local to the town
College player	the college circuit	local in the college circuit
Mia Hamm	the world	global

Now consider a variable's *scope*. If a variable is defined *within* a function, it is a *local* variable because it will only be recognized *within that function*. A variable's scope refers to the largest possible area of recognition. For *local* variables, the scope is usually the function where it was defined.

If a variable is recognized *everywhere*, we call it a *global* variable. All functions recognize a global variable. A global variable must be defined at the *top* of a program, *outside* and *above* all the functions, including the main function. In our example using the struct, we put the definition of the struct before all the function headings to ensure that it would be a *global* variable. We also say that its scope (extent of recognition) is global because it extends throughout the entire program.

```
#include <iostream.h>
#include <string.h>

int x;    // a global variable

struct book   // a global struct ( another global variable)
{
string title;
```

```
string author;
int year;
char discount;
};
```

Notice the variable, *x;* it is also a global variable. All functions will recognize *x* because it has been defined *above* everything. All functions that follow can now use the "book" type. We do not need to redefine it inside of every function that would want to use a "book" type variable.

HINT!

Using the same name for a variable in different functions is okay because these functions are separate, so there will be no confusion for the compiler. In more elaborate programs where functions call back and forth to one another, this may not be the case.

WATCHOUT!

Global variables, although useful, can pose problems for programmers because all functions have access to the global variable <u>at any time</u>. Sometimes the values in a global variable will be inadvertently changed and this may not be what the programmer intended.

SUMMARY

We introduced the *record,* which is a data type designed to hold different data types. Each type within a record is called a *field* of the record. In the programming language C++, a record is known as a "struct." When a struct is defined, it is given a name. This name refers to the new type of variable that has just been created.

Records can be assigned values just like any variable. To assign values to a struct, we assign each of the fields of the struct using the field names.

We introduced an array of structs. This is very useful because we frequently need to hold data for several structs, not just one. Next, the *scope* of a variable was introduced. The scope represents the extent to which a variable is recognized in a program. If a variable is recognized everywhere in a program, we say that the variable is *global*. If it has a limited scope, then the variable is *local*. Variables defined at the top of a program are global because all functions, including the main function, can recognize them.

ANSWERS TO ODD-NUMBERED EXERCISES

10.1 (i)

1.

```
struct movie
{
 string  name;
int year;
double gross;
};
```

3. a "contact_info" type that contains fields to represents a person's home phone number, cell phone number, and email address.

```
struct contact_info
{  string home_no;
   string cell_no;
   string e_mail;
};
```

10.1 (ii)

1. second_individ by the user

```
cout << "Please type the first name."<< endl;
cin >> second_individ.first_name ;
cout << "Please type the last name." << endl;
cin >> second_individ.last_name;
cout<< "Please type the middle initial."<< endl;
cin>> second_individ.mid_init;

struct item
{
string item_type;//pants? shorts? shoes? etc.
```

```
int floor;//floor where item is found
double price;//the price of the item
int percentage;//highest percentage discount allowed on //item
};
```

.

```
struct CD
{
string name;//the name of the CD
string artist;//the artist's name
double price;//price of CD
int year_released;//year CD came out
};
```

10.1.iii

1.

```
item clothing;
clothing.item_type = "jeans";
clothing.floor = 3;
clothing.price = 32.99;
clothing.percentage =  35;
```

3.

```
CD myfavorite;

cout << "Please type the name of the CD."<< endl;
cin >> myfavorite.name;
cout << "Please type the name of the artist."<< endl;
cin >> myfavorite.artist;
cout << "Please type the price of the CD."<< endl;
cin >> myfavorite.price;
cout << "Please type the year the CD was released."<< endl;
cin >> myfavorite.year_released;
```

5. *student* is the name of the new type
"name" is a string
"grade_level" is an int
"year_of_grad" is an int.

7. *book* is the name of the new type

"title" is a string
"author" is a string
"year" is an int
"discount" is a char

CHAPTER

11

Dealing with the Outside World: Files.

IN THIS CHAPTER

- The Text File
- How to Create a Text File
- The Eoln Marker
- The Eof Marker
- The User vs. the File
- Streams of Info/Data

11.1 THE OUTSIDE WORLD FROM A PROGRAM'S PERSPECTIVE

So far we have seen that programs manipulate values, stored in variables. These values have come from either the programmer or the user who interacts with the program. Whenever a program deals with a great quantity of data, the data usually comes from an outside source. (No user wants to sit at the computer typing for hours.)

An example of a program that uses a large quantity of data is a program that prints all the batting averages of all the Major League baseball players in order, from highest average to lowest average. Another example is a program that prints all the names of people with the last name, Smith, who live in the Manhattan borough of New York City. These programs deal with large quantities of data.

A. Smith
Al Smith
Allan Smith
Allan C. Smith
Allen Smith

Large quantities of data (information) can be stored in special files that are called *text files.* Just a couple of examples are the names of all the people in the local phone directory or the names and statistics of all the National Hockey League Players. These files are different from programming files because what they contain is treated only as *text* (keyboard letters and symbols) rather than *commands.* Programming files are different from text files because a *compiler will examine the programming file looking for commands to translate* and then *attempt to execute those commands.*

In this chapter, we will learn how to create a text file. The next step is to learn how to access the contents of the text file. We need to write another program file that does this. There are certain things that are true about text files and the way a programming file "reads" a text file. As we mentioned before, a programming file (a program) generally gets its data from the programmer or the user. This time, we want the program to connect itself to a text file. We want the program to "open" the text file and "read" data from it.

11.2 CREATING A TEXT FILE

The real power of programming comes when you can connect your program to a text file where much data is stored. In our first example, let's create a short text file of our friends' names and phone numbers so we can learn the basics about text file structure. Consider this list of friends with phone numbers:

Carl Brady	555-1234
Marlo Jones	789-0123
Jason Argonaut	888-4567
Jim Collins	456-2345
Jane Austen	234-8765

The text file should be thought of as a large piece of lined paper on which we will write data on each line. Data (information) should be put into the text file with some organization. In this file we have chosen to put a name and then a phone number *on each line*. But first, we need to create the text file and name it appropriately.

When you open a new file and save it for the first time, use some name like

```
"MyFriends.dat".
 ⇓          ⇓
```

file name   extension

We need the extension, ".dat" to indicate that we are writing a *text file*, not a programming file. (Recall that extensions are used to classify files—you have probably already seen .gif, .jpg, .doc files, as examples.)

ORGANIZING A FILE THROUGH MARKERS

When you write a text file (also called a data file), it is important to give it some structure. As far as the compiler is concerned, a text file is just a sequence of letters and symbols. It looks like this:

Carl Brady555-1234 Marlo Jones789-0123Jason Argonaut888-4567Jim Collins . . .

The compiler "does not care" what names you have typed at the keyboard, because it is not looking for programming commands or keywords such as int, main, void, endl, for, while, do, and so on to translate. It will, however, look for markers on the file. Markers are used to separate data. Because data appears in a stream (just like our "cin" stream), the markers are used to provide some organization to what would, otherwise, be a continuous stream. For this reason, the compiler needs to have some organization imposed on the file. The organization in a file will come from the spacing (the blanks that surround a value) and two types of markers that are put onto the text file.

THE EOLN MARKER

The first marker here, we are using a "□," which is an an end of line marker (written "eoln"). This is used to organize the data in the text file. When you write a file and decide what data will be inside of it (e.g., names followed by phone numbers), you then need to put an *eoln* marker to separate each piece of unique information. In this case, we want to write a person's name followed by a phone number on each line. After every new person's name and phone number, we hit the *return* key which causes an *eoln* marker to appear on the file. The file is organized with data followed by these line markers.

THE EOF MARKER

When you finish typing a text file, there is an additional marker automatically put onto the file. It is called an end of file marker ("eof"). We have used this symbol, "•," to represent the marker so you can see where it is placed. This mark is necessary for the compiler or translator that uses the file (Figure 11.1).

Screen View

```
Carl Brady        555-1234□
Marlo Jones       789-0123□
Jason Argonaut    888-4567□
Jim Collins       456-2345□
Jane Austen       234-8765□•
```

Stream View

```
Carl Brady  555-1234□Marlo Jones  789-0123□Jason Argonaut  888-4567□
Jim Collins  456-2345□Jane Austen  234-8765□•
```

FIGURE 11.1 The file is shown as it appears on the screen and as it appears as a stream where each name and phone number is followed by an *eoln* marker (here, a "□") and then at the end of the file, an *eof* marker is used (here, a "•").

11.3 READING FROM A TEXT FILE

Once you have created the text file, it is necessary to think about the next step—reading from the text file. First, we need to examine how text files are "read." A text file behaves like an audio cassette tape. An audio cassette tape can be rewound, fast forwarded, recorded over, and listened to. A text file behaves in almost the same way. When we first use a text file, we have to *open* it and indicate that we are going to *read* from it. *Reading* a text file is like "listening to a tape." Reading means that we will move through the text file, extracting (pulling out) copies of values (like the names and phone numbers in "MyFriends.dat") that are in the file so that we can use them in our program. When we read a text file, we are not altering it or "recording" over it—just "listening" to it.

An analogy to reading a text file is listening to tapes or a music CD. Reading from a text file is almost the same action. The text file already exists, if we created it previously, and now we are using the text file (not changing it) by reading it.

The term *writing* a text file is used to describe creating or changing an existing text file *by sending values out* to the text file. We use the same terminology to describe "burning" a CD—we are *writing* new material *onto* the CD as opposed to just reading (listening to) the CD.

The steps involved in reading and writing text files (data files) are the same for most programming languages. The first step is to "open" the text file and then *indicate* whether you are intending to "read" it or "write" onto it in the program. Before we do these steps, we need to familiarize ourselves with two *streams* that are used with text files. If you recall, streams are like channels through which values can travel or be stored temporarily.

STREAMS USED WITH FILES

Text files are an outside source of information for the programmer. Just as the user (another outside source) can provide values for the programmer, a text file (often called a data file) also provides values. Once a text file has been created, it is necessary to *access* the values from the file so that they can be used in a program.

It is important to familiarize ourselves with the streams associated with a text file because these streams provide us access to the file—whether we intend to "read from" or "write to" the file. In Chapter 2, we talked about the "cin" stream (in the C++ programming language) associated with the keyboard. When you work with files, you will probably have to define a stream that connects to the text file. Recall that the "cin" stream is already being used and is "connected" to the keyboard.

There are two types of streams used with text files: an "in" stream and an "out" stream. In the programming language C++, the "in" stream *type*, called "ifstream," allows data (values) to travel from *the text file* into *the program*. The "out" stream *type*, called "ofstream," allows values to go from *the program* out to *the text file*.

These streams are very similar to the "cin" and "cout" streams we studied previously. If you recall, the "cin" stream is like a channel through which values *come into a program* from data entered at the keyboard. The "cout" stream allows values to go *from the stream out to the screen.*

Some languages do not bother with letting the programmer access the stream directly. For example, in the programming language Pascal, the programmer just uses commands such as "read" and "write," which simplify this process but also make it more difficult to understand how the computer does what it does.

HOW TO ACCESS THE VALUES IN A TEXT FILE

When you begin a program that will use values that come from a text file, you need to indicate to the compiler that your program will be communicating with an outside file. This is done by using an *open* command in the programming language.

We tell the compiler that we will open a file for the purpose of reading from the file. The command *could* look like this:

```
open ("MyFriends.dat");
```

The next step is to declare an "in" stream that will connect to the text file. The "in" stream type is called an "ifstream." (This variable type is found in the header file called *fstream.h.*) Think of this type as short for an in-file-stream. We will call the new stream "file_in." We can call it anything, even "x," but we want to give it a name that reminds us that it is a stream used for *reading from* a text file.

```
ifstream              file_in;
   ⇓                     ⇓
```

type of file stream   name of file stream

Notice that it follows the grammar of declaring a variable of some type.

```
int x;
char initial;
double m;
ifstream  file_in;
```

Once you have declared the stream type, you are ready to use it in the same way you use the "cin" stream, which allows us to get values from the user. We use the same extraction operator that we used with "cin" (>>). We also need to use variable holders for the values that will be pulled out of the stream. In the text file we created previously, we have two strings on each line: a string for the name of our friend and a string to hold a phone number.

```
string name, number; // declaring two strings to hold values from
//the file

file_in >> name >> number ;
 ⇓          ⇓           ⇓
stream name   string name   string name
```

All of these are variable holders.

ACCESSING THE CONTENTS AND DISPLAYING THEM ON THE SCREEN AT THE SAME TIME

The most useful thing to do with a large file is to examine its contents. This can be done by opening a text file for reading and then using the appropriate stream with a loop that will allow you to get copies of all the values from the file.

We will use the previous example and expand it in two ways. The first statement we will add is a "cout" statement so that we can see the values that came from the text file (They were copied into our variables, "name" and "number.") and display them on the screen.

```
string name, number; // declaring two strings to hold values from
//the file

file_in >> name >> number ;
cout << name << "    " << number << endl;
//this statement allows us to view name's and number's values.
```

The example only lets us see the *first two* values in the text file—"Carl Brady" and "555-1234." Every time you use the extraction operator with the file stream, you are reading from the file, which is like "playing a tape." If you want to read the entire file (like playing a tape all the way to the end), you will have to use the extraction operator repeatedly until everything has been copied out of the file.

HINT!

Reading from a file does not change a file. It is similar to listening to a tape or playing a CD. The file is not changed.

If you want to see all the contents of the text file, you will need to use a loop so that you can get all five lines of data from the file.

```
string name, number;
int y;

for ( y = 1 ; y <= 5; y= y + 1 )
{
file_in >> name >> number ;
cout << name << "    " << number << endl;
}
```

Each time this loop spins, a new name and number get copied into the two variable holders. Then the "cout" line allows the program to print the values on the screen. Think of the "file_in" stream with the ">>" operator as something that pulls copies of values out of the file and then puts them into the variable holders, in this case, *name* and *number*. The next line uses the "cout" stream to send those variables' values to the screen.

HINT

When using a loop with the *same* two holders (as in our previous example), the values in the variables will constantly change as the loop spins.

11.4 SOME ELEMENTARY COMMANDS USED WITH FILES

The best analogy for a text file (also known as a data file) is a VCR cassette tape, which you can play, record over, rewind, and fast forward. The text file behaves similarly to the cassette tape. For these reasons, there are a small group of commands (or functions) associated with using files in any programming language. Let's look at these commands so that you will be familiar with the types of operations that you can perform on a file.

HOW TO "REWIND" A FILE

When you *open* a file, there is an imaginary pointer that moves back to the beginning of the file. This "pointer" keeps track of where you are in the file as the file is being read. It's like the counter on your VCR, which goes to zero after you rewind a tape. When you *read* from a file using the file stream we mentioned earlier, the pointer will move through the file as you read from the file (Figure 11.2).

```
⇓
Carl Brady   555-1234 □ Marlo Jones   789-0123 □ Jason Argonaut   888-4567 □
Jim Collins  456-2345 □ Jane Austen   234-8765 □ •

                        ⇓
Carl Brady   555-1234 □ Marlo Jones   789-0123 □ Jason Argonaut   888-4567 □
Jim Collins  456-2345 □ Jane Austen   234-8765 □ •

                                                ⇓
Carl Brady   555-1234 □ Marlo Jones   789-0123 □ Jason Argonaut   888-4567 □
Jim Collins  456-2345 □ Jane Austen   234-8765 □ •


Carl Brady   555-1234 □ Marlo Jones   789-0123 □ Jason Argonaut   888-4567 □
                                                ⇓
Jim Collins  456-2345 □ Jane Austen   234-8765 □ •
```

FIGURE 11.2 A file is shown with a pointer ("⇓") in different places in a file.

WRITING VALUES TO A FILE

If you want to add another number or name to a file, there are two ways to do this. The first way is to type more data *directly into the file* just as we did when we first created the file "MyFriends.dat." Just add more names and numbers onto the bottom of the list you already have.

```
Carl Brady              555-1234
Marlo Jones             789-0123
Jason Argonaut          888-4567
Jim Collins             456-2345
Jane Austen             234-8765 //This was where the file used
                                 //to end.
William Shakespeare     789-1564 //Now it ends here.
```

The other option is to *open* the file and use the stream that allows data to go out to a file. This type of stream is called an "ofstream" stream. (You could think of it as an "out" to file stream.) Let's declare a stream of the "ofstream" type.

```
ofstream              file_out;
   ⇓                     ⇓
type of file stream   name of file stream
```

Now we can use the insertion operator, <<, that we used with the "cout" stream from Chapter 4. The insertion operator will allow the two variables' values to be sent out to the stream and then to the file.

```
string A = "hello!";
int B = 78;
file_out<< A << B;
```

The text file will look like this:

hello!78

If we want to put *spaces* between values in a text file, we will have to put spaces *into the stream,* that is, the channel to the text file. Look at this next example where we insert spaces between the values that are going to the stream.

```
file_out<< A << "    " << B;
```

The text file will look like this:

hello!   78

To keep the file structured with values on separate "lines," we will *put a line marker into the stream with the endl command* so that a line marker goes onto the file, itself.

```
file_out<< A << "    " << B << endl;
```

HINT!

The return key is used to put a line marker directly onto the text file; the *endl* command must be used to put a line marker into a stream that will send its values to a file.

ADDING VALUES TO THE END OF A FILE

If we want to get to the end of a text file, we will have to read it all the way to the end—like "playing an audio tape" to the end. Some languages provide a command that allows us to go to the end of a file so that we can add more values there. This command is called "append." When you "append" something, you are adding it onto the end of something. (An *append*ix appears *at the end* of a book.)

Let's start by opening the file, "MyFriends.dat," and appending another name and number to it.

```
open ("MyFriends.dat");
append ("MyFriends.dat");
string A = "Mark Holden";
string B = "456-1234";
ofstream file_out;
file_out << A << "     "  << B  << endl;
```

Now the file should look like this:

Carl Brady	555-1234
Marlo Jones	789-0123
Jason Argonaut	888-4567
Jim Collins	456-2345
Jane Austen	234-8765
William Shakespeare	789-1564
Mark Holden	456-1234

CLOSING A FILE

The last thing to do when you are finished with a file is to close the file. (Just like closing a book when you are finished reading it).

```
close ("MyFriends.dat");
```

Every language has its own commands for using files. It is important to see how each new language handles files. You need to check to see how you can open the file, write to it, read from it, and then append, if possible. The last thing is to remember to close the file.

ON THE CD-ROM!

A program in the programming language C++ that creates a text file. Another program that reads data from this text file and displays it on the screen.

SUMMARY

Text files (also called data files) are outside sources of data (files of data) that can be introduced into a program. Text files can be one or more lines of data or text and do not have any programming commands. The big advantage of a text file is not having to rely on the user or the programmer as sources of large quantities of data. Phone directories, lists of students at schools and universities, and statistics for members of professional sports are just a few examples of large quantities of data. When you name a file, you should use the ".dat" extension, which indicates that the file type is not a programming file.

Text files have *two different kinds of markers* used to organize information within the text file: the *eoln* marker and the *eof* marker. When you create a text file directly, you need to hit the return key to place an *eoln* marker onto the file. An *eof* marker is automatically placed onto a file after it has been created.

Reading from a text file allows copies of values to come out of a file for use in a program. Writing to a text file sends values out to a file through a program. To perform either of these tasks, you need to use the two streams associated with text files. In the C++ programming language, these streams are called *ofstreams* and *ifstreams* (*out* to *file* streams and *in* from *file* streams). They work in the same way as the "cout" and "cin" streams we discussed earlier.

Basic commands associated with files include opening, closing, and appending to text files. There is an imaginary pointer (like a counter for a tape machine) that sits above a file and keeps track of where you left off in the file. When you open a file, the pointer moves to the beginning of the text file. When you append to a file, the pointer automatically goes to the end of the file. Closing the file will cause the file to be closed, just like closing a book.

CHAPTER

12

Pointers: Who Is Looking at Whom?

IN THIS CHAPTER

- Static Variables
- Dynamic Variables
- Introduction of Pointer Variables
- How Pointer Variables Work
- Why Use Pointer Variables?

12.1 STATIC VS. DYNAMIC VARIABLES

There are two kinds of variables and, before we understand the basic difference between them, we need to consider an analogy. Imagine you are taking an around-the-world trip where you will visit different countries in different climates. You expect to spend the first part of the trip in northern countries where the climate is cold. For the second part of the trip, you will be visiting southern countries where the *weather* is warmer. Strategically, you will pack half your luggage with clothing for the cold weather and the rest of your luggage with clothing for the warmer climate in southern countries.

After you spend the first part of the trip in the cold weather, you plan to send that luggage home by air mail so you don't have to continue to carry it around with you—as, literally, excess baggage. Now you can travel more easily carrying only what you need for the remainder of the trip.

Variables can be likened to the baggage on a trip. You are stuck with some variables for the entire program. You can't get rid of them even if their purpose was long over after the first few minutes of running a long program. They are called *static* variables because they exist for the duration of a program. Other variables, called *dynamic* variables, are convenient in the way that they can be "unloaded" at any point during the execution of a program.

STATIC AND DYNAMIC VARIABLES AND THEIR ROLE IN MEMORY

There are two classifications of variables beyond the types we mentioned earlier in the book. These classifications describe *how variables use memory.* The first classification of variables is called *static. Static variables* exist for the duration of a program. Once you declare a static variable, *it occupies a spot in the memory* of the computer *until the program ends.* All the variables we have studied so far have been static variables.

Dynamic variables behave differently. They can be *declared during the execution* of a program. What that means is they can *come into existence* during the running of the program *if they are needed.* They *can also be destroyed* during the running of the program. When a dynamic variable is destroyed, *the space it occupied* in memory *is released* back to the computer. This is important if you are running a program that uses huge amounts of memory and you don't want to waste any of it on variables that no longer serve any purpose as the program continues to run.

12.2 THE POINTER VARIABLE: A DYNAMIC VARIABLE

In this section we examine the *pointer variable,* which is a dynamic variable. A pointer variable (or pointer, for short) is a variable that *holds an address* rather than a *value.* Think of a variable as a container. Normally, a variable holds a value. Regular variables look like holders with values inside (Figure 12.1).

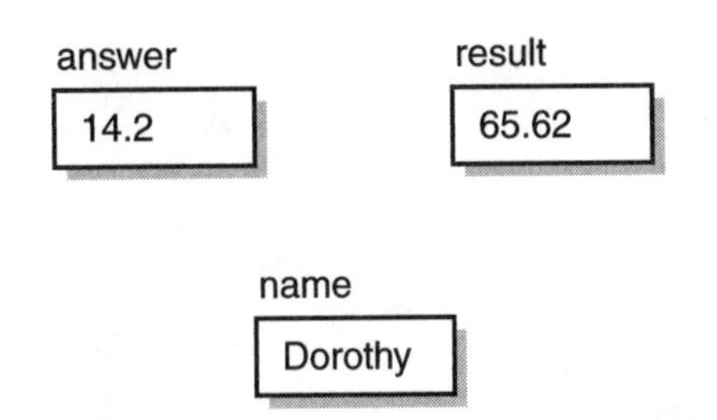

FIGURE 12.1 Three *static* variables are shown with their values inside.

Pointer variables, unlike static variables, *do not contain values. They contain addresses*, which look like a combination of letters and numbers (Figure 12.2).

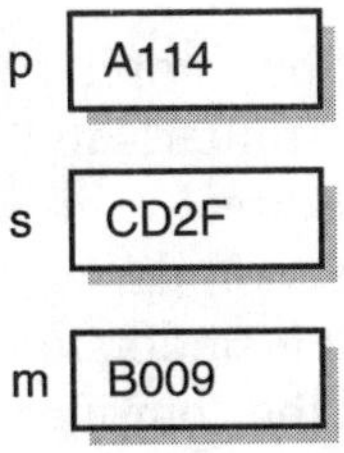

p A114

s CD2F

m B009

FIGURE 12.2 Three *pointer* variables are shown with addresses inside of them.

HOW A POINTER (VARIABLE) WORKS

Now that we know that pointers contain addresses, the question is, "How do they work?" The first step is to *get the address* that is contained in the pointer. The next step is to *go to that address* (where another variable is found). Then *you can go inside* of the other variable to get *its* value. Because the pointer contains an address, it really "points" to another place in memory. At that other place in memory, you will find a variable that contains a value (Figure 12.3).

12.3 OPERATIONS ASSOCIATED WITH POINTERS

There are different tasks or operations associated with pointers regardless of the language you are using. Once you understand the concepts behind each of these tasks, you will be ready to learn the different syntax rules particular to each language. Let's examine each of these tasks.

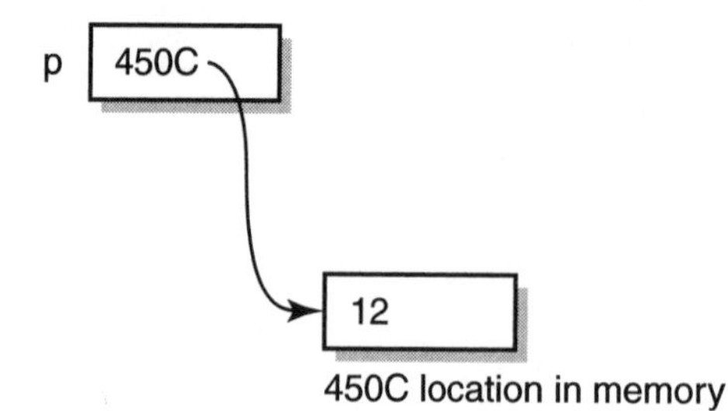

FIGURE 12.3 A pointer is shown with its address. Another variable containing a value is shown at that address.

DECLARING A POINTER

When you declare a pointer, you are telling the compiler to treat the variable in a different way. The compiler will "understand" that it will be dealing with memory addresses rather than direct values. When you declare a pointer variable, you need to tell the compiler what type of value the pointer will ultimately be "looking at." It is understood that a pointer will contain an address, but we need to tell the compiler an address for some type of variable. Will it contain an address to an integer, a character, a string, or a record/structure? These are important concerns for the compiler.

Just like the declaration of a static variable, we first state the type of variable followed by the pointer's name. The only difference between static variable declarations and pointer variable declarations is the use of an intervening symbol to indicate that it is a pointer. In the C++ language, we use an asterisk (*) between the pointer's name and the variable type it will point at.

```
int * p ; // p will "look at" an integer
// p will contain the address of an integer variable.
// p, itself, is not an integer!
```

HINT!

The pointer variable has a name, but the variable it looks at does not always have a name. The other variable is known as "the variable the pointer looks at" (Figure 12.4).

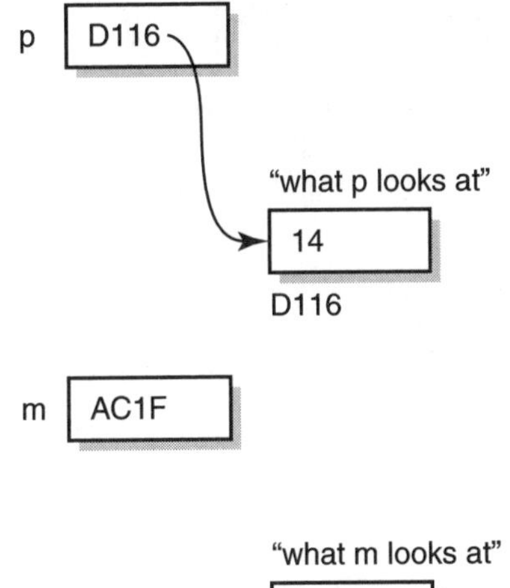

FIGURE 12.4 Two pointer variables are shown with their own variable names and the addresses they contain. At each address, another variable is shown containing a value. These variables are labeled "what p looks at" and "what m looks at."

ALLOCATING MEMORY FOR A POINTER

The term *allocate* means to set aside for use. If you "allocate" funds for a project, such as building a bridge, for example, you set aside funds to build the bridge. When you allocate memory, you are reserving memory for use by the pointer. The command to set aside memory is the "new" command. This command is used with the type of variable the pointer will look at. The compiler will look for an available space, get the address of that space and assign that address to the pointer. The pointer *now contains that address.* Consider this example.

```
int * p; // p is declared as a pointer variable

p = new int; // memory is set aside (allocated)  for p
```

We don't have any control over the address that the compiler finds in memory for the pointer. The compiler will look for a place in memory that is open or unoccupied. It will then assign the address of that spot to the pointer. The next step is to assign a value to that address according to the type used in the declaration. Once we have allocated memory for a pointer, then we want to assign it.

ASSIGNING THE VARIABLE AT WHICH THE POINTER "LOOKS"

There are two kinds of assignment for pointer variables, depending on how you want the pointer to be used. The first type of assignment is when you assign the variable "the pointer looks at." This assignment is just like the other assignment statements we have used previously. The only difference with this kind of assignment is that we need to clearly tell the compiler that we wish to assign "the variable the pointer looks at"—not the pointer itself.

Let's call the variable the pointer "looks at," the *referred* variable.

Let's look at how we name this other variable. If we use the correct syntax for naming the variable, then we can use that name in the assignment statement to make the statement work. When calling this other variable, we use the asterisk symbol (*) again but we put it *after* the pointer name. Consider these three statements used with the pointer called "p."

```
int * p; // The compiler is informed that p is a pointer.
p = new int; // memory is allocated for p.
p* = 6 ; // "what p looks at" is assigned the value, 6.
```

Here are some other examples of pointers being declared, then assigned, after memory has been allocated to them.

```
char * g;  // g declared as a pointer to a character
g = new char; // memory is allocated to g
g* = 'm';  // "what g looks at" is given the value, 'm'.

double * num;
num = new double;
num* = 45.65;
```

WATCHOUT!

It is very important to use the asterisk symbol properly. In declaration statements, it is used on one side of the pointer name. In assignment statements, it is used on the other side of the pointer.

EXERCISES

For each of the following, declare a pointer that "looks at" the given type.

1. An int
2. A char
3. A double

For each of the following pointers, assign the given value to where the pointer "looks."

4. The pointer in problem one, the value 35
5. The pointer in problem two, the value 'p'
6. The pointer in problem three, the value –87.11

EXAMPLES TO DISTINGUISH BETWEEN THE POINTER AND THE REFERRED VARIABLE

In the following examples, consider each value in the variable. Either the value is an *address*, contained in the pointer, or it is a *value* of the variable type *referred to* by the pointer. Also note the names of each variable, both the pointer names and the variables referred to by the pointer names (Figure 12.5).

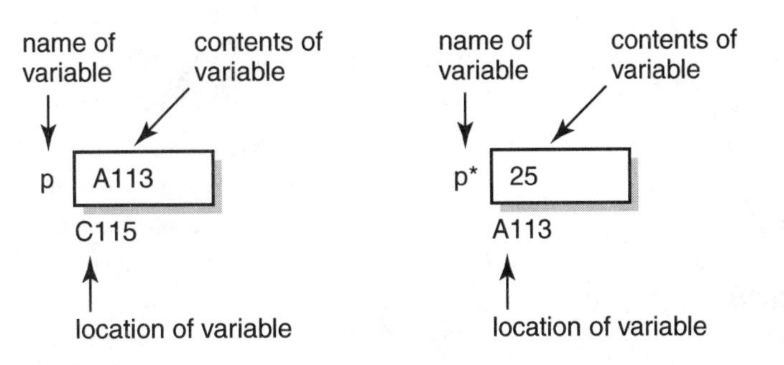

FIGURE 12.5 Each variable, whether a static variable or a dynamic (pointer) variable, is located at an address and contains something either an address or a value.

In this section, we will examine statements that either use the pointer or the "referred" variable. Because pointers contain addresses, it is sometimes necessary to consider the address contained in a pointer. In the first example, we want to check to see whether two different pointers are "looking at" the same variable. The easiest way to do this is to compare the addresses contained in the pointers. Let's start by declaring two pointers and assigning their "referred" variables.

```
int * p;// both p and m are declared as pointers
int * m;/* The compiler now "knows" they will contain addresses. */

p = new int; m = new int;//memory is set aside for both
p* = 15;// the "referred" vars are assigned
m* = 34;
```

Next we want to compare the pointers themselves, using a relational operator. Let's assume that p's address is AB14 and m's address is C12A. The next statement will compare their addresses to see if they are the same.

```
//addresses contained in both p and m are compared
if ( p == m )
cout << "These pointers are looking at the same variable.";
```

The compiler will check to see if the relational expression, AB14 == C12A is true, which it is not. If we use the following statement with our pointer variables, the relational expression will be true because AB14 does not equal C12A.

```
if ( p != m )
cout << "These pointers are not looking at the same variable.";
```

Let's look at some other examples where we use either the pointer or its "referred" variable. The statements are shown at left and the values printed on the screen are shown at right. Notice that for the last two examples, we do not know what the user will type—the '?' indicates this fact.

EXAMPLES

```
cout << p << endl;                    AB14
cout << p* << endl;          15

cout << m << endl;                    C12A
cout << m* << endl;          34

//Letting the user assign p*
cin >> p* ;                           ?
```

```
//Letting the user assign m*
cin >> m*;                        ?
```

EXERCISES

Identify whether the variable is a pointer variable or its "referred" variable. (All variables refer to the int type.)

1. snaker*
2. p
3. n*

Complete each assignment statement appropriately.

4. snaker* =
5. p =
6. n* =

HOW TO ASSIGN THE POINTER ANOTHER POINTER

So far we have assigned only the variables "referred to" by the pointers. It is also possible to assign the pointer itself. *We can assign the pointer an address contained in another pointer.* This is equivalent to making two pointers *look at the same variable.* One pointer's address will be copied into another pointer's variable after executing these statements.

```
int * p;  // declaring a pointer to an int
int * snaker;  // declaring a pointer to an int

snaker = new int; // allocating memory for the snaker
p = snaker; // p will "look at" the same variable as snaker
// because p now has the same address that snaker contains.
```

HINT!

As we said before, we have no control over the address the compiler assigns to a pointer when we allocate (set aside) memory to the pointer.

DEALLOCATING MEMORY

When we speak of "deallocating" memory, we really mean that we are releasing occupied space back to the compiler for its use. If you recall, it's like sending the luggage home during the long trip. How it works is best explained in this way. When we allocate (set aside) memory for a pointer through the "new" command, the compiler finds a block of memory both *suitable for the variable* type and *unoccupied,* and assigns that address to the pointer. The pointer is the gateway to that other variable's contents. This variable, the one we call "the variable looked at by the pointer," is the *referred* variable. If we need this variable for the duration of our program, it behaves much like the static variables we studied previously.

However, if we find that we need more memory and this referred variable is no longer useful, then we can release this memory back to the compiler. Releasing the memory is called "deallocating memory." We do this through the "delete" command.

```
//p is a pointer variable
delete p; // p's address is given back to the compiler
```

A variable containing some value occupies memory space. When the pointer is used with the delete command, the address of the referred variable is returned to the compiler for future use.

HINT!

Deallocating memory is the same as releasing the memory occupied by the referred variable. The address is given back to the compiler for future use.

SUMMARY

We introduced static variables and dynamic variables. Static variables exist for the duration of a program whereas dynamic variables do not. The pointer variable is a dynamic variable that contains an address to a location in memory. At that other address is a variable, called the referred variable or "the variable the pointer looks at." This other variable contains a value. The referred variable is the variable looked at by the pointer because the pointer contains the address for this variable.

We gave different examples of the syntax for assigning the referred variable of a pointer. You have no control over the address the compiler chooses to set aside for a pointer. The pointer is given a spot in memory—suitable for the type of variable the pointer "points to" or "looks at." Allocating memory for a pointer (setting aside memory) is done through the "new" command. Likewise, releasing memory once referred to by a pointer is called deallocating memory and this is executed with the "delete" command.

The asterisk (*) symbol is used in pointer syntax. When you declare a pointer variable, the asterisk is used in the declaration to indicate that a pointer is being declared rather than a static variable. (All the variables we have worked with so far have been static variables.)

ANSWERS TO ODD-NUMBERED EXERCISES

12.3 (i)

1. int * x ;
3. double * result;
5. using char * m, for example, m* = 'p';

12.3 (ii)

1. the referred variable
3. the referred variable
5. p = snaker;

CHAPTER

13

Searching: Did You Find It Yet?

IN THIS CHAPTER

- Searching through a List, Using an Array
- The Linear Search: Examining One Member at a Time
- Writing a Loop to Do This Work
- Inserting a Relational Expression to Find Something
- The Binary Search: Explained with a Guessing Game
- Defining an Index: The Left, Middle, and Right Indices
- Using Indices in the Binary Search

13.1 SEARCHING

This chapter covers the topic of *searching* (i.e., looking for a value amid a large group of values.) Generally we search though an array for a value. The array, you might recall, is a data structure that can hold a large number of values. Consider this array, called a "list" of values assigned by the user.

Array Member	*Containing Value*
`list[0]`	22
`list[1]`	–5
`list[2]`	6
`list[3]`	10
`list[4]`	0
`list[5]`	–37
`list[6]`	14
`list[7]`	2
`list[8]`	141
`list[9]`	59
`list[10]`	46

The programmer does not know which values are in the array because the user thought of the values that went into each slot.

SEARCHING FOR A VALUE IN THE ARRAY

Let's say the programmer wants to know whether the value –37 was typed by the user. (He does not yet know it is in member list[5].) We can write a fragment of code to search through the array, *looking at each member* of the array to see whether it is the value, –37.

Is it in list[0]? Is it in list[1]? The programmer writes code to check each slot of the array until he finds the slot that contains –37. He will eventually find that –37 is in the 6th slot of the array. We will look at the code that accomplishes this in the next section.

SEARCHING FOR A VALUE NOT IN THE ARRAY

Had the programmer wanted to know whether the value 100 were in the array, he would have been disappointed because it is not there. The programmer would write code to look at each member of the array to see whether 100 was in one of the slots of the array. List[0] does not contain 100, nor does list[1], list[2], or list[3].

The programmer must write code so that the computer looks at each member. Eventually he will get to the last few members of the array—list[8], list[9], and list[10]. He'll confirm that 100 is not in any of them.

HINT!

The only way to confirm that a value is not found in an array is to check *every* slot of the array first.

13.2 DEVELOPING THE LINEAR SEARCH ALGORITHM

The programmer needs to methodically move through the array, examining one member at a time. The *linear search* is the name for this *method* of searching—looking for a value by *examining one member at a time in an array*.

HOW TO EXAMINE A MEMBER

The easiest way to examine array members is to use a relational expression with each member of the array and the value for which we are searching. Let's call the holder for this value "Number." In our previous example, "Number" contained –37. We need to examine each array member. We will use a relational expression with the equality symbol, ==. This is the general form of the expression we will use.

Array Member == Number
Does the value in the Array Member equal the value in Number?

The programmer must have the computer check to see whether this relational expression is true. He'll program the computer to check each slot of the array until he finds one (a slot or member) that contains –37.

Does list[0] == Number?	Value of Relational Expression
22 == –37	false
Does list[1] == Number?	
–5 == –37	false
Does list[2] == Number?	
6 == –37	false
Does list[3] == Number?	
10 == –37	false
Does list[4] == Number?	
0 == –37	false
Does list[5] == Number?	
–37 == –37	true

Once the relational expression is true, he can stop searching. He has found the value in Member 5 of the array "list."

WRITING A LOOP TO DO THIS WORK

A pattern is developing in the previous questions containing relational expressions. In each expression we are examining:

list[slot #] == Number

Let's use a variable to represent the slot in the array. Let's declare it an integer called *x*.

```
int x;
x = 0;
```

```
            Does list[x] == Number?
                       ⇓
            Does list[0] == Number?
                  ⇓              ⇓
                  22     ==    –37              false
```

If *x* changes in value and becomes 1, then we can check the next slot (member) of the array.

```
x = x + 1; // x is now 1
```

```
            Does list[x] == Number?
                       ⇓
            Does list[1] == Number?
                  ⇓              ⇓
                  –5     ==    –37              false
```

If *x* changes again, we can check the next slot (member) of the array.

```
x = x + 1; // x is now 2
```

```
            Does list[x] == Number?
                       ⇓
            Does list[2] == Number?
                  ⇓              ⇓
                  6      ==    –37              false
```

We are ready to develop a loop that will move through each member of the array.

```
for (int x = 0;  x <= 10; x = x + 1)
{
// does list[x] == Number?

}
```

DEVELOPING AN IF...STATEMENT WITH THE RELATIONAL EXPRESSION

We want to write an *if...statement* that contains the relational expression. First we need to decide what to do when we find the member that contains the same value that is in *Number*. Let's allow the user to know *where* we found the value by printing a message on the screen.

```
if  (list[x] == Number)
cout << Number << "is in the array at position"<< x << endl;
```

The next step is for us to put this statement *inside of* the loop we just constructed. Now the entire fragment will look like the following block of code:

```
for (int x = 0;  x <= 10; x = x + 1)
{
if  (list[x] == Number)
cout << Number << "is in the array at position"<< x << " ." <<
endl;
// this is a very long output statement
}
```

13.3 THE BINARY SEARCH

There is another kind of search that is very efficient in finding a number in an array. The *binary search* can be performed on an array that is *already in* some kind of *order*. Before we examine the binary search, we need to talk about an array that is "in order."

ASCENDING ORDER

The word "ascend" means "to go up"—like ascending a mountain. The phrase "ascending order" means that numbers *increase* as you move through a list. Let's look at a list that is in ascending order. Notice that you can skip a lot of numbers in a list and still be in "ascending" order.

3 4 6 17 21 24 32 43

These numbers increase as you move through the list from left to right. Let's construct an array that contains these numbers. We'll use an array of 8 members called *"group."*

```
int group[8];

group[0] = 3;
group[1] = 4;
group[2] = 6;
group[3] = 17;
group[4] = 21;
group[5] = 24;
group[6] = 32;
group[7] = 43;
```

Once an array is "in order," we can begin to use a *binary search* on the array when we are looking for a number. We will examine the case where the number we are searching for is in the list.

WHAT IS A BINARY SEARCH?

We know that a search is a method by which we methodically "look for" a number. We have to remember that we can't always "see what is inside" an array—especially if the user loaded the numbers through an input statement such as this:

```
for (int x = 0;  x <= 8; x = x + 1)
{
cout << "Please type a number." << endl;
cin >> group[x];
}
```

We need to keep in mind that we do not know the contents of an array. For the binary search to work, we do need to know that the numbers are *in order.*

The word "binary" is used to indicate that, as the search is being performed, there are two halves of the array that we will examine. The binary search is a method that involves cutting the array in half, figuratively (Figure 13.1).

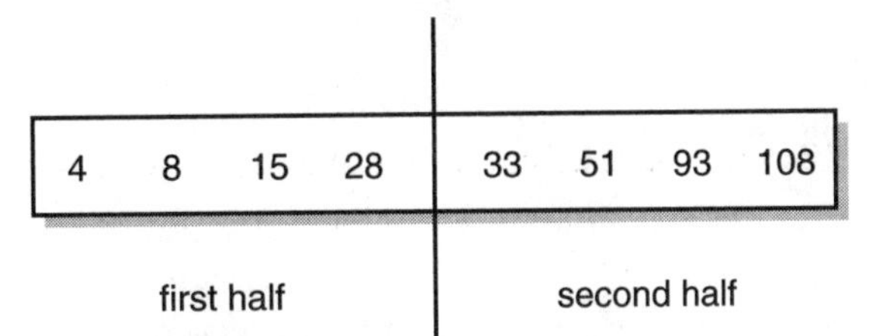

FIGURE 13.1 An array is shown cut into two *halves*. This is the reason we call the search a *binary* search.

UNDERSTANDING THE BINARY SEARCH

Now imagine that you were playing a game with a friend who was fond of numbers and had lots of free time on her hands. The game is for you to guess the number that your friend is thinking of. Instead of telling you whether you have guessed correctly, your friend will give you one hint for each guess. She will respond whether your guess is higher or lower than the correct answer. You and your friend, Eloise, have this conversation:

Eloise: "I am thinking of a number between 1 and 100. Guess what it is." (Eloise is thinking of 34.)
You: Is it 50?

Eloise: "No, it's lower than 50."
You: "Is it 25?"

Eloise: "No, it's higher than 25."
You : "Is it 37?"

Eloise: "No, it's lower than 37."
You: "Is it 31?"

Eloise: "No, it's higher than 31."
You: "Is it 34?"

Eloise: "That's it!"

It took you five guesses to figure out what number Eloise was thinking of. (If you had used the linear search it would have taken 34 guesses!) There is also a pattern to the guesses you made. Each time Eloise responded with a hint—saying that the answer was either *higher* or *lower* than the guess—you, correctly no longer considered the numbers that were *out of range*. Another interesting pattern that occurred in the game is that your guess was always in the *middle* of the part of the list being examined.

EXAMINING THE MIDDLE OF A LIST

The guessing game is a good way to see how the binary search works. You always start by picking a number *in the middle* of your list as *your first guess*. Then, if you find out that your guess was wrong, you need to know whether the correct answer is higher or lower than your guess. Once you know that, you can *cross off* the part of the list that is *out of range*.

Think of the first guess you made in the example. You guessed the number 50 and Eloise's response was that the answer is lower than 50. Now you can throw out the numbers 50, 51, 52, ..., 99, 100, because the answer must be less than 50. Here is the advantage of the binary search at work. You are able to cut out an entire half of the array so that your next guess will come from a *smaller* section of the array (Figure 13.2).

Guessing the middle of the list is very efficient because it maximizes the numbers you can eliminate. If you were to make a guess that was not in the middle, it would be very inefficient. Consider what would happen if you played the same game with Eloise but this time you had the following conversation:

Eloise: "I am thinking of a number between 1 and 100. Guess what it is." (Eloise is thinking of 34.)
You: "Is it 1?"

Eloise: "No, it's higher than 1."
You: "Is it 2?"

Eloise: "No, it's higher than 2."
You: "Is it 3?"

Eloise: "No, it's higher than 3."
You: "Is it 4?"

Eloise: (exasperated!) "No, it's higher than 4."

By now Eloise has figured out that you will probably continue your pattern of guessing the next biggest number. (You are actually performing a linear search by going through the list of members, one at a time.) We might now begin to see the advantage of guessing a number in the middle of a list. It allows us to *throw out a large section* (half) of the list that you know *does not contain the answer*.

1 2 3 ... 47 48 49 50 51 52 ... 98 99 100

Is it 50?
No, it's lower.

1 2 3 ... 47 48 49 ~~50 51 52 ... 98 99 100~~

⇓

1 2 3 ...23 24 25 26 27 ... 47 48 49

Is it 25?
No, it's higher.

~~1 2 3 ...23 24 25~~ 26 27 ... 47 48 49

⇓

26 27 ... 36 37 38 ... 48 49

Is it 37?
No, it's lower.

26 27 ... 36 ~~37 38 ... 48 49~~

⇓

26 27 28 29 30 31 32 33 34 35 36

Is it 31?
No, it's higher.

~~25 26 27 28 29 30 31~~ 32 33 34 35 36

⇓

32 33 34 35 36

Is it 34?
YES!

FIGURE 13.2 A list is shown with each guess displayed and a successive part of the array eliminated. Notice how the actual part of the list that you examine gradually becomes smaller and smaller.

Most people (who have not studied the binary search) would randomly guess numbers because that would seem like a better alternative than guessing the numbers one at a time from left to right. You might get lucky or you might not. For that reason, we guess a number in the *middle* of the list.

HINT!

Remember that the binary search can only work on numbers that are arranged *in order*.

MOVING TO THE RIGHT OR LEFT

When you make a guess that is incorrect, you need to know whether your *guess was lower or higher than the answer.* This is where the binary search saves you time—by eliminating half of the list. When you make a comparison between your guess and the answer, you need a relational expression with the "greater than" (>) operator:

```
if (my_guess > answer)
/* I'll throw out all the numbers greater than (to the right of)
  my_guess. */
```

This is what happened when we first guessed the number 50. Eloise responded that the answer was "lower than 50." That is, our *guess* was *too high* a number. Our guess (50) was *greater than* the answer (34). As a result, we started to consider only the numbers that are *lower* than 50 or to the *left* of 50—as if these numbers were arranged in a list from left to right.

In another guess we made, we guessed 25, and Eloise responded that the answer was "higher than 25." In this case, our guess was too *low* a number. So the answer had to be *higher* than 25, or to the *right* of 25 if we lined up the numbers from left to right in a list. Here, a relational expression using the "less than" (<) operator would have been appropriate:

```
if (my_guess < answer )
/* I'll throw out all the numbers less than (to the left of)
  my_guess. */
```

Every time you lose half of the array, you save some time in your searching process. In an *ordered* list of the numbers 1 through 1000, your first guess using the binary search would be the number 500. If the guess is not correct you can immediately throw out 500 members of the list (either those numbers less than 500 or greater than 500). What a great timesaver when you are searching for a number!

HINT!

Remember that with a list written in ascending order, *numbers to the left* of other numbers are *less than* those numbers. *Numbers to the right* of other numbers are *greater than* those numbers.

13.4 HOW THE BINARY SEARCH WORKS WITH AN ARRAY

Now that we have studied the basics of the binary search, we need to consider how programmers use it in programs. We mentioned earlier in the chapter that we generally search an array because the array is a data structure that can hold quite a few numbers. Let's start to look at the binary search as a way of finding a number in an array whose values are in order.

FINDING THE MIDDLE OF THE ARRAY

To find the middle of the array, we need to find the middle *member* of the array. If the array has 8 members, then the 4th member is the middle member. If the array has 50 members, then the 25th member is the middle member.

As you continue to use the binary search, you need to make adjustments in how you find the middle member. For example, if you eliminate the first 500 members of an array that has 1000 members, you will need to find the middle of the subsection of the array that begins with the 501st member (Slot 500) and ends at the 1000th member (Slot 999).

The subsection you are examining:
500 501 502 503 . . . 748 749 750 751 . . . 997 998 999
the middle of that subsection of the array

A vertical representation of the array:

```
list[500]
list[501]
list[502]
list[503]
.
.
.
list[748]
list[749]  //  the middle of the array subsection
list[750]
list[751]
.
.
.
list[997]
list[998]
list[999]
```

When we found the middle of the arrays we mentioned previously, we used the size of the array to find the middle. In the array with 8 members, we know that half the number of members would be 4, so Slot 4 is the middle of the array. (It is actually in group[3] because we began the array at `group[0]`.)

An *index* is a slot of an array. All arrays in the programming language C++ begin with the index, 0. We call this a *left index* because it represents the leftmost slot of an array when it is viewed from left to right. Likewise, the last slot is the *right index*. In the case of an array with 8 members, the right index would be 7. In an array with 500 members, the right index would be 499. In an array with 1000 members, the right index would be 999.

HINT!

You always need to remember that (in C, C++, Java) we begin arrays with the 0 index.

As you eliminate halves of the array through the binary search, a "new middle" is found by using the left and right indices of the subsection of the array. Let's take our previous example—where we were examining the right half of the array (we'll call it "list") with 1000 members.

500 501 502 503 . . . 748 749 750 751 . . . 997 998 999

With this subsection, the left index is 500 and the right index is 999. We find the middle index, the index of the middle member, by taking the *average* of the left and right indices.

middle index = (*left index* + *right index*) / 2

Let's see how the middle index is assigned by doing this average. The left index (500) + the right index (999) is 1499. Dividing 1499 by 2 gives us the number 749.5. Because slots of the array are always *integers*, division with integers causes the excess (0.5) part of the number to be dropped. So the new middle of the array is Slot 749.

Let's find the middle index of the array *group*. The left index is 0 and the right index is 7 so the middle index is found by taking the average of the two indices. (0 + 7 is 7 and 7/2 is 3—because we eliminate the 0.5.)

HOW TO "ELIMINATE" HALF OF THE ARRAY

The binary search is efficient because we are able to "throw out" half of the array each time we get the answer that the number we are searching for is "higher than" or "lower than" our guess. Let's practice the search on the example of the array *group*.

Let's start by searching for a number that is in the list. Here is the list again to remind us of the values that are in it. Let's ask the user for a number to search for in the array. We will store it in a variable called "answer."

```
int group[8];

group[0] = 3;
group[1] = 4;
group[2] = 6;
group[3] = 17;
group[4] = 21;
group[5] = 24;
group[6] = 32;
group[7] = 43;
cout << "Please type a number and I will tell you if it is in the
list. " << endl;
cin >> answer;
```

Using the "guessing game" we played earlier, let's assume the user types a number that is in the list—such as 4—it will represent the "answer." We'll now play the guessing game with the computer, instead of with our friend, Eloise. We (the programmer) play the game by the rules of the binary search, which means we always make a guess—a value found in the middle slot of the array. (Our guess, in the guessing game, is *always* the value found in the middle slot of the array.)

Programmer: "I am looking for a number in my array *group*. (The user picked the value 4.) I'll first check the number found in the middle of the array—at group[3]—the value *17.* "

Computer: "No, your guess, 17, is higher than the answer (4). Examine the left half of the array—that is, all the slots of the array *to the left* of the middle slot ([3])."

Programmer: "O.K. I'll focus on members with indices, 0, 1, and 2, because that is all that remains to be examined according to the binary search. Is the value in Slot 1?"

Computer: "Yes! How did you get that so fast?"

Programmer: "I chose the middle index of the subsection of the array. The left index was 0 and the right index was 2 so the average—the middle index—is 1. The value in the middle slot (`group[1]`) matches the value in the variable `answer`."

When we adjust the indices for a subsection of the array, we are giving new values to both the left and right indices. Consider what happened in the game played between the programmer and the computer. After the "first guess" was made, the subsection of the array was the left half of the array. We get this left half of the array by looking at the members from the *left index*, 0, to the *right index*, which is the *last index before the middle index*—here, it is 2. After we establish the values of the left index (0) and the right index (2), we can use their average—(0 + 2)/2—to find the value of the middle index—1. The answer is always found in the middle slot of some section of the array. Every time you eliminate half the members, that subsection gets smaller and easier to work with.

ON THE CD-ROM!

The complete binary search is shown in a program in the C++ programming language.

EXERCISES

Please give the middle index for each of these arrays:

1. int list[86];
2. double collection[142];
3. int block[2020];

Now give the middle index for a subsection of each array that has each of the following indices:

4. left index 0 right index 69
5. left index 43 right index 85
6. left index 1010 right index 2019

SUMMARY

We began the chapter by defining what it means to search through a list for some value. We usually search through an array because the array can hold so many values. A linear search is a searching method by which we examine a list (an array) one member at a time. Because we repeatedly examine members of a list, we can use a loop to do this work.

We also use a relational expression to find a value in an array. The relational expression is put inside of the loop so that all members can be examined for the value. It is possible that the value is not in the list.

The binary search represents a search method that is very different from the linear search. It is more efficient than the linear search but it can only be used on an ordered array—one whose values are listed in increasing size—from left to right. We call this order *ascending order* because the numbers ascend as we move through the list.

The binary search can best be explained through a guessing game played between two people. Later in the chapter, the game was played between the programmer and the computer as an illustration of the binary search at work.

The binary search uses the middle of the list as its first guess in the guessing game. If the guess is incorrect, the next guess will be found in a subsection of the array—found either to the left or right of that initial guess.

By focusing on one half of the array before making the second guess, we have eliminated half of the original list. For this reason, the binary search is very efficient.

An index is a slot in the array. The left index is, at first, the slot numbered 0. The right index will be the slot of the last member of the array. To find the middle index, we take the average of the values of the left and right indices. (If there is any fractional leftover, we discard it because an index must be an integer.)

ANSWERS TO ODD-NUMBERED EXERCISES

13.4

1. 42 because (0 + 85)/2 is 42.5
3. 1009 because (0 + 2019) /2 is 1009.5
5. 64 because (43 + 85) is 128 and 128/2 is 64

CHAPTER

14

Let's Put Things in Order: Sorting

IN THIS CHAPTER

- Alphabetization Defined
- Ascending vs. Descending Order
- The Selection Sort
- Finding the Minimum
- The Merge Sort

14.1 HOW WE ARRANGE DATA—SORTING

When we *arrange* data in order, we *sort* data. It is important to be able to arrange a list of words in alphabetical order or a list of numbers in increasing order of size. If you had a list of lottery numbers, you would want to be able to arrange them so that you could easily look up a certain number in a list. There are many ways to *sort* data and as you study the topic of programming, you will learn these different sorting methods. In this chapter, we will discuss the broad topic of sorting and then study two particular sorts so that you can see how a sorting algorithm works.

Think of each sort as a *recipe* for arranging data in order. Each sort has its own name, depending on the particular way data is arranged in order.

The ability to sort is an important task when you are dealing with large quantities of data. The most obvious kind of arrangement of data would be the alphabetization of strings—putting words in alphabetical order. Names (strings) are alphabetized all the time. Look at this list of names that are *not* in alphabetical order:

John
Nicola
Marie
Susan
Brian
Bridget
Margaret
Nick
Bernie
Al

If we alphabetize this list, we see that Al should be first and then we will examine the names that begin with 'B.' After handling those names, we can look to the next letter in sequence, the 'J.' Then we find that the names beginning with 'M' should be placed before those beginning with 'N' and, last, we follow with the name "Susan."

Al
Bernie
Brian
Bridget
John
Margaret
Marie
Nick
Nicola
Susan

If you stop for a moment and analyze the work you did to alphabetize this list, you might not think it was much work at all. After all, you can see that "Al" is first because the name begins with an 'A.' Next, you might admit that it required a little work to arrange the three names that began with 'B' in order. (You need to look at the second and third letters to see which name is first, second, and third among them.) Now stop to consider how a machine with no intelligence would go about organizing data in some kind of order.

Here we begin the topic of sorting—arranging data in order. First we will introduce two kinds of order—ascending and descending order. In the last chapter, we examined ascending order. We will be sorting data in ascending order in this chapter.

14.2 ASCENDING VS. DESCENDING ORDER

Ascending order is the order we introduced in Chapter 13 when we mentioned the requirement that a binary search can only be performed on an array that is in order. Ascending order means that numbers in the array *increase in value* as you move through the array. For example, as you move from Member 3 to Member 4, you expect the value found at Member 4 to be greater than that found at Member 3. Look at this example of a list of numbers in ascending order. (The numbers increase in value as you move from left to right.)

–8 –1 2 5 12 18 19 32 41 45

If an array were used to contain these values, the array's first holder slot would hold –8 and the second holder would contain –1, and so on.

–8	–1	2	5	12	18	19	32	41	45
list[0]	list[1]	list[2]	list[3]	list[4]	list[5]	list[6]	list[7]	list[8]	list[9]

Descending order is just the opposite—the values found at the front of the array (or on the left) are bigger in value than those that follow (on the right). Look at the same list of values, this time listed in *descending* order.

45 41 32 19 18 12 5 2 –1 –8

This list is in descending order because every value (to the left) is bigger than every value to the right. Now let's take this list and put the values into an array.

```
int list[10];

list[0] = 45;
list[1] = 41;
list[2] = 32;
list[3] = 19;
list[4] = 18;
list[5] = 12;
list[6] = 5;
```

```
list[7] = 2;
list[8] = -1;
list[9] = -8;
```

As you move *down* through the list, the values become *smaller*. This array is in *descending order*. We will examine a few different sorts to see the basic method by which numbers are arranged in order.

14.3 A SORT: THE SELECTION SORT

For a computer to sort (organize) data, it must follow a very specific set of instructions. There are different kinds of sorts, depending on the method used. Each sort is named for the method it employs. The *selection sort* is a good sort to begin your study. As an analogy, let's start by looking at an example of a selection sort performed on a group of people at a party and then we'll examine the sort applied to a list of numbers in an array.

EXAMPLE

AN ANALOGY: THE PARTYGOERS

Imagine a group of people attending a party. Someone who is not a part of the party plans to make a list of the youngest to oldest people in attendance. We will call this person the organizer. The organizer enters the room where the party is being held and asks the first person he meets how old he or she is. The organizer continues and asks every other person how old each is. One by one he tries to find someone younger than the first person he asked. If he finds someone younger, he considers that person *the youngest* now. He continues to ask each person his or her age. When he finishes querying everyone, he retains the name of the person who was the youngest.

The organizer asks that person to leave the room after he writes down the name first on his list. Returning to the room, he asks the remaining people, one at a time, their ages. He will find the youngest person in that group and ask that individual to leave the room. He writes down the name of the person leaving.

The organizer now holds a list of two people—the youngest person's name is at the top and the next youngest person is listed second. The party is not as big as it was before because *two people have left*. As the organizer polls the remaining group to find the next youngest person, he writes down that person's name and asks him or her to leave. Now the party is missing three people and is, thus, an even smaller group. He now has a list of three people—the youngest, second youngest, and third youngest of all the people that went to the party.

The algorithm we are developing is to *select* the youngest in a group and put that person's name at the top of a list. Then we select the youngest person (really the second youngest) from the slightly smaller group (the youngest is missing because we asked her or him to leave) and follow the same procedure; write down the name and ask the person to leave.

The organizer keeps a list of each "new" youngest person found at the party and also asks that individual to leave once he has written down the name. In this way, he can examine a smaller group each time to find the "new" youngest person.

The list he keeps is growing. It is a list of the youngest through oldest people who attended the party. By repeatedly selecting the youngest person from the group, he is organizing the names of all the people who attended the party in order from youngest to oldest.

TWO PARTS TO THE SELECTION SORT

The first part of the selection sort is to notice that we are *selecting* the youngest from a group and putting that person's name at the top of the list. The second part of the sort is to recognize that we are examining a smaller and smaller group each time (Figure 14.1).

Each sort you will study has some structure to its method. The selection sort selects the smallest number from a group and puts it at the top of a list. This number is removed from the group so that when you find the smallest in the remaining members of the group, you will be able to find the next smallest, and so on.

14.4 FINDING THE MINIMUM

Before we start to look at the selection sort performed on an array of numbers, we need to look at the part of the sort where we "found the youngest person in the room." This requires some work.

Finding the *youngest* person in a room is an analogy for finding the *minimum* in a list of numbers. The minimum is the *smallest value* in a list.

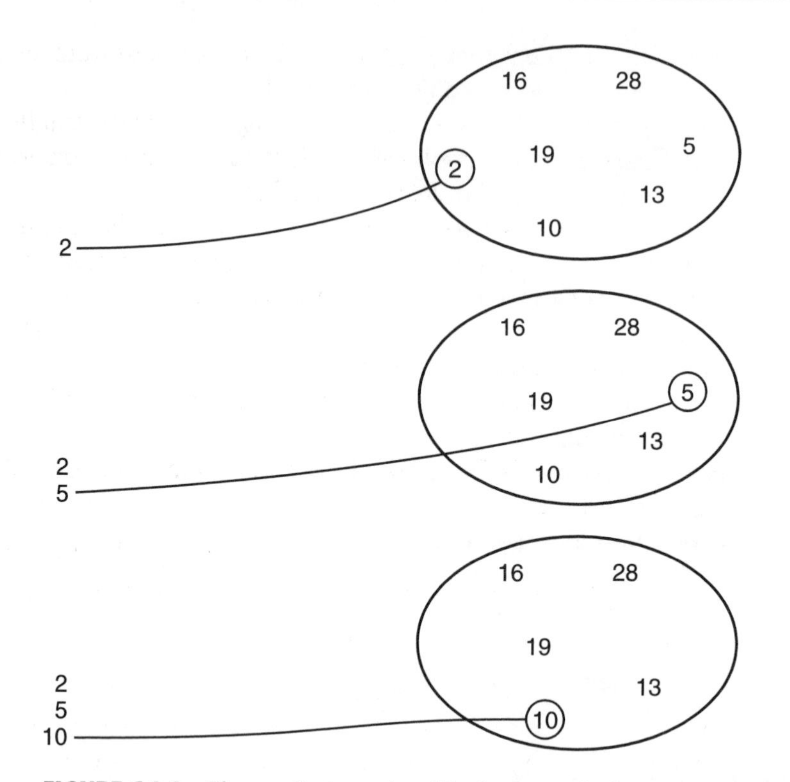

FIGURE 14.1 The smallest number (2) of a group is shown outside the group at the top of a list. Then the group is shown *without* that member. Next, the smallest (5) is chosen from the group and is written second in the list. The third time, the smallest (10) is pulled from the group and written third in the list. This is how the selection sort works.

When you find the minimum of a list, you are really finding the smallest value in a list of numbers. Let's examine this list of 6 numbers:

14 32 5 12 61 7

As you look at the list, you can immediately see that 5 is the smallest number in the list. The task is more difficult for a computer. It has to be programmed to "find" this minimum. Let's look at what the computer might "think," in order to better understand how finding the minimum works.

Computer: "Let's assume that 14 is the smallest number. Let's get the next number from the list—32.

Is 32 less than 14? No. So 14 is still the smallest. Good. (less work for me)"

Computer: "Let's assume that 14 is the smallest number. Let's get the next number from the list—5.
Is 5 less than 14? Yes. I guess 5 is the smallest now.

Computer: "Let's assume that 5 is the smallest number. Let's get the next number from the list—12.
Is 12 less than 5? No. So 5 is still the smallest.

Computer: "Let's assume that 5 is the smallest number. Let's get the next number from the list—61.
Is 61 less than 5? No. 5 is still the smallest. (It has lasted longer than 14.)

Computer: "Let's assume that 5 is the smallest number. Let's get the next number from the list—7.
Is 7 less than 5? No. So 5 is the smallest value in the entire list.

Let's do this exercise a second time showing at least two replacements of the minimum value as you move through the following list:

21 16 25 8 19 4 1

Looking at this list, you can see that 1 is the smallest. But consider the work the computer would do if confronted by such a list (an *array*)! Remember that the computer cannot "see" all the numbers in the list at once.

Computer: "Let's assume that 21 is the smallest number. We'll get the next number from the list—16.
Is 16 smaller than 21? 16 is now the smallest in the list. (21 didn't last long!)

Computer: "Let's assume that 16 is the smallest number. We'll get the next number from the list—25.
Is 25 less than 16? No. So 16 is still the minimum.

Computer: "Let's assume that 16 is the smallest number. We'll get the next number from the list—8.
Is 8 less than 16? Yes. So 8 will replace 16 as the new minimum. (Fair enough.)

Computer: "Let's assume that 8 is the smallest number. We'll get the next number from the list—19.
Is 19 less than 8? No. 8 is still the minimum.

Computer: "Let's assume that 8 is the smallest number. We'll get the next number from the list—4.
Is 4 less than 8? Yes. So 4 is the new minimum. (The minimum has changed three times.)

Computer: "Let's assume that 4 is the smallest number. We'll get the next number from the list—1.
Is 1 less than 4? Yes. So the minimum has changed again.

After the computer has "exhausted" the list by examining all the members, the minimum of the entire list is the value that *remains* in the minimum holder (variable) at the end of the algorithm.

When you program the computer to find the minimum, you first *establish a value* for a variable called "minimum." Using the *first value in the array* is an efficient choice. Consider this fragment of code:

```
int collection[7];

collection[0] = 21;
collection[1] = 16;
collection[2] = 25;
collection[3] = 8;
collection[4] = 19;
collection[5] = 4;
collection[6] = 1;

int minimum = collection[0]; /* the first value in the array is
assumed to be the smallest since we have seen no other values. */
```

Now that the minimum has been assigned a value, the computer will need to look at all the other values in the array—one at a time—to make comparisons with the value in the "minimum" holder. If any "better value" is found (a lower value), then the minimum will need to be assigned that *lower* value. Look at this example of a statement that would cause the minimum to be replaced by a new *lower* value:

```
if (collection[3] < minimum)
minimum = collection[3];
```

In the comparison between the values in collection[3] and minimum, the computer will evaluate whether it is true that collection[3] contains a value that is lower than the value contained in "minimum." If the value of that relational expression is true, then "minimum" will be reassigned the value that is contained in collection[3]. The old value in "minimum" will be "replaced."

Now look at all the work the computer does to find the minimum for this array of 7 members. Here are six statements that will look for a "better minimum" (a lower value) than the minimum that is established initially with the value in the first slot of the array.

```
if (collection[1] < minimum)
minimum = collection[1];
if (collection[2] < minimum)
minimum = collection[2];
if (collection[3] < minimum)
minimum = collection[3];
if (collection[4] < minimum)
minimum = collection[4];
if (collection[5] < minimum)
minimum = collection[5];
if (collection[6] < minimum)
minimum = collection[6];
```

This is how the computer finds the lowest value in the entire list; it methodically looks for a lower value *in each* of the slots of the array. The *most important thing* to notice, however, is that *whenever the computer finds a lower value, it immediately* replaces the value in *minimum.*

You might have noticed that the preceding statements seem repetitive and they are. We need to use a loop to save some time in our programming. Let's use a *for loop* to "hit" all the members of the array.

```
for (int x = 0; x < 7; x = x + 1)
{
if (collection[x] < minimum)
minimum = collection[x];
}
```

Now we are ready to see the entire block of code that allows the computer to find the minimum from an array of numbers. This block of code will work on any array; you only need to adjust the *for loop* so that it will spin the appropriate number of times.

```
minimum = collection[0];//setting the initial (first) value
// for the minimum variable.
for (int x = 0; x < 7; x = x + 1)
{
if (collection[x] < minimum)
minimum = collection[x];
}
```

```
cout << "The smallest value in the array is "<< minimum
     << " ." << endl;
// remember the cout line can wrap to the next line.
```

HINT!

Remember to assign the value of the minimum variable the first member of the array, SLOT 0.

HINT!

To find the minimum, the computer needs to constantly compare a *new array slot* with the *most recent value in the minimum* variable. It is possible for the value in the minimum variable to change several times before the *for loop* ends.

ON THE CD-ROM!

A program that shows the complete selection sort

14.5 MERGE SORT: A VERY FAST SORT

The next sort we will discuss is called *merge sort,* which involves bringing together two halves of an array. The way it sorts numbers in order is quite different from the "selection" sort we just studied. For this reason, it is an interesting sort to study.

Let's take a list of numbers from an array to see how "merge sort" works on the list to organize it into ascending order (numbers increasing as you move from left to right or from top to bottom). We will name each slot of the list underneath the number.

32	12	5	18	31	4	25	7
[0]	[1]	[2]	[3]	[4]	[5]	[6]	[7]

HOW MERGE SORT WORKS

Consider the array we just presented, cut in half. It would look like this:

32 12 5 18 31 4 25 7

⇓ ⇓

32 12 5 18 31 4 25 7

one half another half

If we cut it again in half, it will create four quarters:

32 12 5 18 31 4 25 7

one half another half

⇓ ⇓ ⇓ ⇓

32 12 5 18 31 4 25 7

first quarter second quarter third quarter fourth quarter

Each piece of the array is cut into halves (again), creating eight individual elements:

32 12 5 18 31 4 25 7

first quarter second quarter third quarter fourth quarter

⇓ ⇓ ⇓ ⇓ ⇓ ⇓ ⇓ ⇓

32 12 5 18 31 4 25 7

Initially, the array of 8 elements is viewed as one array. After the first "halving," the eight-member array is viewed as two separate four-member arrays. The next cut in each half produces four quarters. The next "halving" of each of those quarters produces the eight individual elements. Consider how the array has been broken into halves repeatedly (Figure 14.2).

Now let's do this exercise again, but this time we will focus on the entire left half of the array—that is, *the first four elements*. We will "ignore" the right side of the array for now:

32 12 5 18 31 4 25 7

one half ignore this half, for now

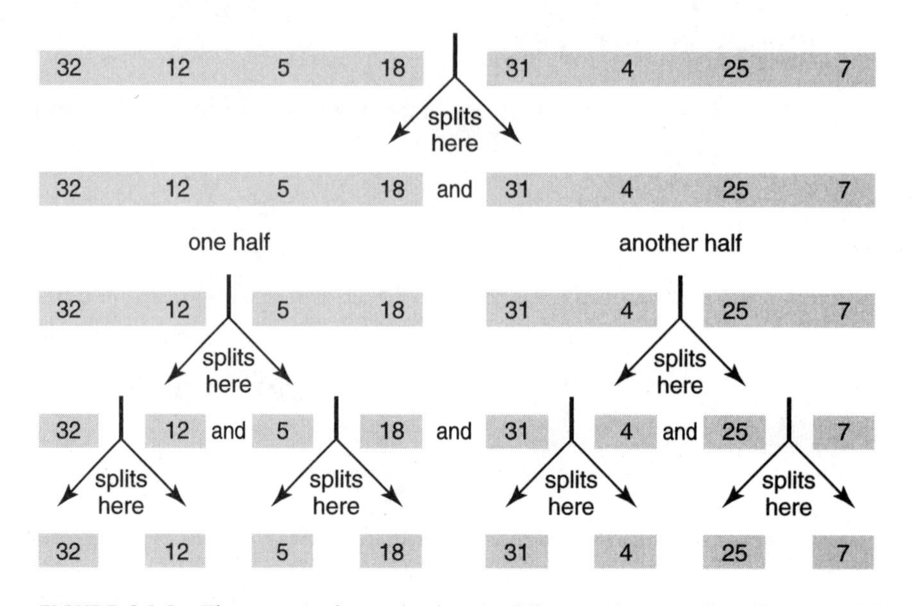

FIGURE 14.2 The array is shown broken in different places until each piece of the array is a single element.

If we cut the left half "in half," we will have two quarters on the left side:

32 12 5 18 31 4 25 7
first quarter second quarter third quarter fourth quarter
IGNORE THIS HALF

Each of the left two quarters is then cut in half. Now there are *four* individual elements on the left side of the array while the right half is ignored.

32 12 5 18 31 4 25 7
four individual elements ignore this half

The left half of the array is ready to be "reassembled" into the proper order. It is important to notice that as we reassemble the left half of the array, the right half of the array remains out of order. Merge sort is programmed to work on one half of the array at a time.

PUTTING THE ARRAY INTO ORDER

Let's consider where we left off in the array. The first four elements need to be "reassembled" or merged into the proper order. The first rule of merging is that you can only merge with the element that you most recently split from. What does that mean with our list? If you look back at the numbers, 32 and 12 can be merged because they were "split apart"; but 12 and 5 cannot be merged because they did not come from the same "quarter" of the array. Likewise, 5 and 18 can be *merged*. Let's merge each of these pairs into the proper order.

32 12 will become

12 32

The two elements have been properly merged and are in order. Next, the 5 and 18 can be merged:

5 18 will remain

5 18

Now look at the left half of the array. It is still not in order but it has undergone some ordering. Each quarter is in order.

12 32 5 18

Now these two quarters of the array can be brought together (merged) into a left half of the array. The 5 will be first, followed by the 12, then the 18, and followed by the 32.

5 12 18 32

Now the array (still undergoing the merge sort) looks like the following:

5 12 18 32 31 4 25 7

left half in order this half was ignored

FOCUSING ON THE RIGHT HALF

Let's do the same thing to the right half as was done to the left half of the array. (We will ignore the left half for now.) The *right* half, just like the *left* half, will be split into two quarters and then into four elements.

5 12 18 32 31 4 25 7

5 12 18 32 31 4 25 7

IGNORE THIS HALF the right half

If we cut the right half "in half", we will have two quarters on the right side:

5	12	18	32	31	4	25	7
first quarter		second quarter		third quarter		fourth quarter	
IGNORE THIS HALF							

5	12	18	32	31	4	25	7
IGNORE THIS HALF				four individual elements			

The last four elements should now be "reassembled" or merged into the proper order. We need to remember that we can only merge with the element that we most recently split from. The numbers 31 and 4 will be merged because they were split apart in an earlier step. The numbers 25 and 7 can be merged. Let's merge each of these pairs into the proper order.

31 4 will become

4 31

The two elements have been properly merged and are in order. Next, the 25 and 7 can be merged:

25 7 will become

7 25

Now look at the *right* half of the array. It is still not in order but it has undergone some ordering.

4 31 7 25

Now these two quarters of the array can be brought together (merged) into an ordered right half of the array. The 4 will be first, followed by the 7, then the 25, and followed by the 31.

4 7 25 31

Now the array has two halves (a left half and a right half) that are ready for a final merge. Each half looks like the following:

5	12	18	32	4	7	25	31
left half in order				right half in order			

The final step of merge sort is to merge these two "ordered" halves. The result of this merging is an array in ascending order (Figure 14.3).

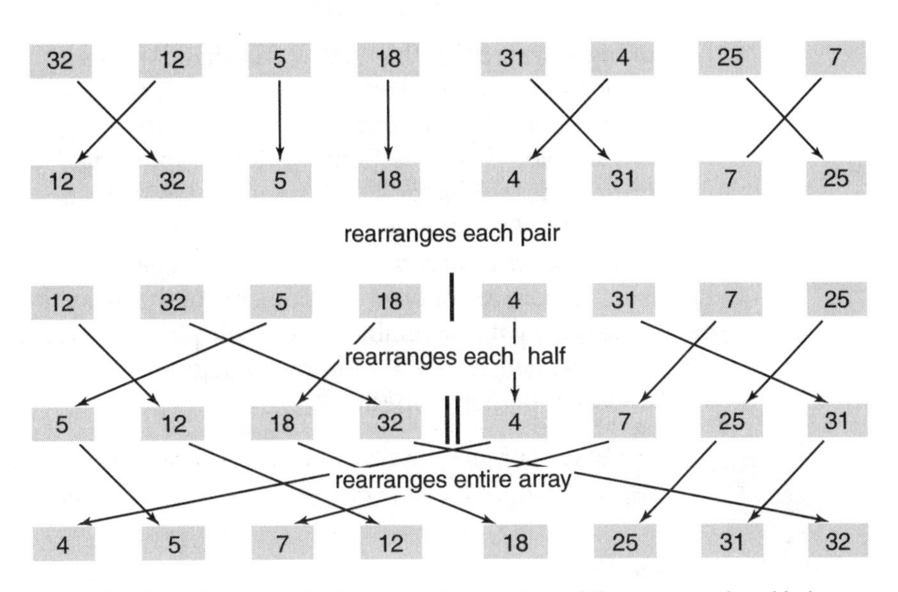

FIGURE 14.3 The array is shown as the merging of these two ordered halves. Each member from each half goes into its proper position.

The merge sort works by a process that allows an array to be cut into halves *repeatedly* until individual elements are left. Then those members are brought together, each half with its other half into the correct order. Eventually the entire array is in order.

The programming code to execute the merge sort involves a process called *recursion,* which we will study in the next chapter.

SUMMARY

Sorting means arranging data in order. Arranging a list of names in alphabetical order is an example of *sorting*. Sorting a list of numbers means arranging them in *ascending* or *descending* order. Ascending order arranges smaller numbers before larger numbers in an array. Descending order puts numbers that are greater in size first, followed by smaller numbers.

Next we covered the *selection sort,* which relies on the ability to find the minimum in a list. The selection sort was explained through an analogy of people attending a party. Then we looked at the selection sort applied to a list of numbers.

Next we defined the *minimum* of a list. It is the *smallest number* in a list. To find the smallest number, the first number from the array (slot [0]) is assigned to a variable called the "minimum." Then the algorithm works by looking at the rest of the numbers in the list—one at a time—through the use of a loop. As you look at each number, you *test whether that number is smaller than* the number that is presently assigned to the "minimum" variable. When the loop is done "spinning," the value in the "minimum" variable is the smallest number in the list.

The *merge sort* was the second *sort* we examined. It works very differently from the *selection sort*. The merge sort repeatedly cuts an array into halves until there are only single elements of the array. Then those members of the array are assembled into their proper order, two pieces at a time. The sort gets its name from the "merging" that occurs as the pieces are put back together properly.

CHAPTER

15

Recursion: Calling Yourself Over and Over Again

IN THIS CHAPTER

- Examples of Recursion
- Definition of Recursion
- Key Features of Recursion

15.1 RECURSION—WHAT IS IT?

Recursion is a particular process involved in solving a problem. Essentially the process involves solving a simpler problem of the same type as the original problem we were asked to solve. Let's first look at some examples outside of the context of programming to understand this process.

EXAMPLE

Let's say we have a vocabulary word and we do not know its meaning. Our second impulse would be to look the word up in the dictionary. (Our first impulse would be to ask someone the meaning!)

Take the word "obstreperous." If you look up obstreperous in the dictionary, it might say something such as "recalcitrant or willful." When you read this meaning, at first you will be happy that there is a word you understand—"willful"—but then disappointed that you need to look up another word—"recalcitrant." So you next look up the word "recalcitrant" to see what that means. This time the dictionary says something such as "defiant." At this point, you might think you know what the word "defiant" means, but you need to check it to be sure. So once again you consult the dictionary and you find that "defiant" means "unruly." The last word, "unruly,"—you are almost sure of the meaning, but you look it up to be certain. Here the dictionary lists " stubborn" (Figure 15.1).

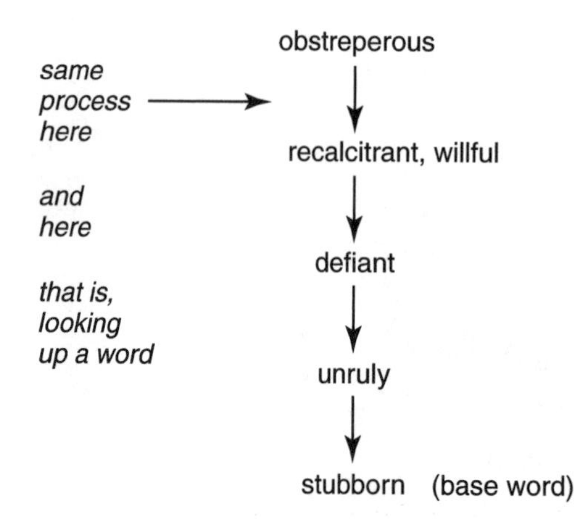

FIGURE 15.1 A word "obstreperous" is shown with other words used to explain its meaning. Each word below another word contributes to the understanding of the meaning of the original word.

So the word "obstreperous" at the base of all this work means something like "stubborn" with some other shades of meaning picked up from the words along the way. Our process of discovering the meaning of this word has been a *recursive* process. We chose a method to solve our problem—finding the meaning of a word—by using the same process, looking up other words in a dictionary. Each time we looked up a new word, we came closer to understanding the meaning of the original word.

Recursion is this process of redoing a problem but in a simpler context than the original problem we did. Consider this next example, which uses recursion and involves numbers.

EXAMPLE

It is often easier to see recursion in a numeric example than one that does not involve numbers. To understand this example, we first need to define the *factorial* of a number. The *factorial* of a number is a product of the number and all the numbers less than it—all the way down to 1.

The factorial of 6 is the answer you get from multiplying 6 by all the numbers less than it—5, 4, 3, 2, and 1. Recall that multiplication is represented by the asterisk, *.

```
6 * 5 * 4 * 3 * 2 * 1
\ /
 30 * 4 * 3 * 2 * 1
   \ /
    120 * 3 * 2 * 1
      \ /
       360 * 2 * 1
         \ /
          720 * 1
          \ /
          720
```

So the answer to the factorial of 6 is 720.

Factorials quickly produce massive numbers because you are multiplying so many numbers, usually, to get an answer. The factorial of a number is represented by an exclamation point (!). The previous example looks like this symbolically: 6 !

Before we get to the recursion involved in the problem, let's redo the problem with some creativity. Let's start with the symbol for a factorial and use it in writing the multiplication for the factorial.

6! = 6 * **5 * 4 * 3 * 2 * 1**
  another factorial is here

Looking at the boldface part of the statement above is another factorial. It is called the factorial of 5, 5 !. Let's rewrite the previous statement using that information.

6 ! = 6 * 5 !

The previous statement means that the factorial of 6 is found by multiplying the number 6 by the factorial of 5. If we only knew the factorial of 5, we could finish this problem right away. So the solution to finding the factorial of 6 rests on our ability to find the factorial of 5 and then to take that answer and just multiply it by the number 6. This way of solving the problem is a *recursive* method. We are solving our problem by solving a *smaller case* of the same type of problem.

Now consider the separate problem of finding the answer *to the factorial of* 5. Consider what the factorial of 5 looks like as a product of numbers.

5 ! = 5 * 4 * 3 * 2 * 1
another factorial is here

Again, just as in the previous example, we can use another factorial to represent part of the work needed to get the answer to the factorial of 5. Let's rewrite the problem using that factorial.

5 ! = 5 * 4 !

And so the process continues like this. We find the factorial of 4 by finding the factorial of 3 and then multiplying that answer by 4.

4 ! = 4 * 3 !

The last two lines of work involve finding the factorial of 3 and the factorial of 2. (The factorial of 1 is the value 1.) Consider these last two lines of computation.

3 ! = 3 * 2 !
2 ! = 2 * 1

Each line of computation involved finding the answer to a similar—but smaller—factorial problem. The important point to notice is that you are finding the answer to the problem using a *smaller case* of the same problem. The factorial of 6 used the answer from the smaller case, the factorial of 5. Another point to notice is that this process eventually stops. If it did not stop, then the recursive process would continue forever.

15.2 TWO FEATURES OF RECURSION

To understand the topic of recursion more comprehensively, we need to look at key aspects of a recursive solution (i.e., what constitutes a recursive solution to a problem). We will also examine how our previous problems (the word in the dictionary and the factorial problem) modeled these aspects. Let's examine the key features of a recursive problem.

1. A smaller problem of the same type exists.
2. Eventually, you reach a bottom case or "stopping state."

Recursion would be useless if it continued indefinitely. We would never arrive at an answer if that were the case. In both examples we completed in the previous section, both of these features were present in our solution of the problem.

When we looked up the word "obstreperous" in the dictionary, we eventually found a simpler word whose meaning we did understand. The simpler word, "unruly" is a "smaller case" of the same type of problem. (The word is not as difficult as the original word.) The stopping state for the word problem is *finding a word whose meaning we do know* so we can stop looking up words in the dictionary.

With the factorial example, the smaller case of the same type of problem is using a *factorial* of a number *less than the number* we started with. (The factorial of 5 is a smaller case than the factorial of 6.) The stopping state is reaching a factorial whose answer we do know—the factorial of 2 because it is the product of 2 * 1.

Infinite recursion is a problem that arises when one of these parts is not satisfied by the recursive design we make. For example, if we never arrived at a word we understood in the dictionary we would always be looking up new words.

USING FUNCTIONS IN RECURSION

To understand how recursion works, we need to look at how functions are the bedrock of recursion on a computer. Imagine if we defined a function called "look_up." This function would look up any word that we did not understand. First, we would call on it and send in the word "obstreperous" from our previous example. Then we would want to call the function again, but this time send in the word "recalcitrant." We would want to call the function a third time, sending in the next word "defiant."

The last time we call the function, we send in the word "unruly," where the meaning we get is clear to us so we can *stop calling the function* "look_up."

The process of looking up one word and then another and another until we get a word whose meaning we know shows that we need to make several calls (4 calls, in fact) to this function (Figure 15.2).

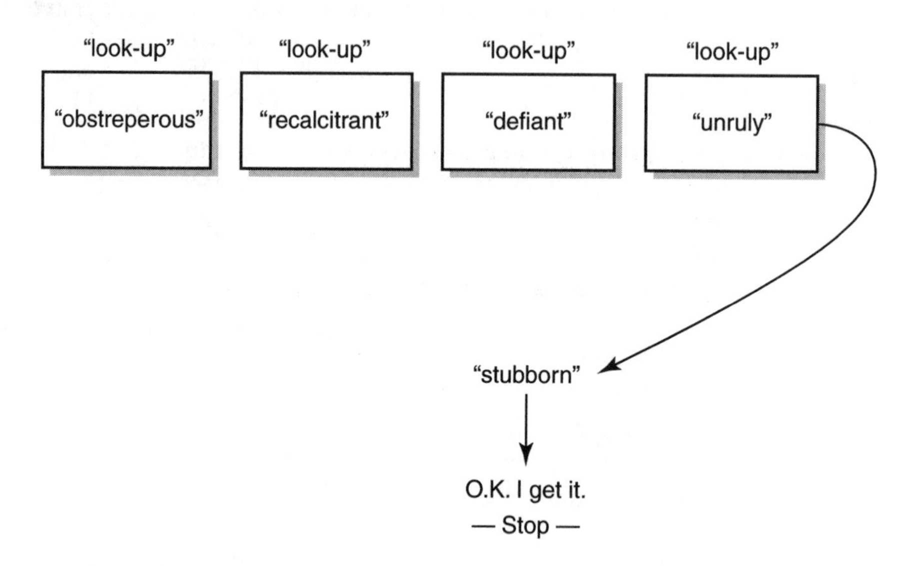

FIGURE 15.2 A chain of calls is shown with a new word sent to the function each time it is called.

ONE FUNCTION CALLS ITSELF

The reality behind these repeated calls to one function is that the calling *occurs within the function* itself. Now what does that mean? We have always talked about the main program (the main function) *making the calls to* the function. In reality, any function can make a call to another function. In the case of recursion, the function calls itself. Let's see what this does.

In the "look_up" example, the main function calls the "look_up" function and sends in the word "obstreperous." Once the "look_up" function starts its work—looking up a word in a dictionary—it finds another word it doesn't know. So it leaves in the middle of its own function and calls "look_up" again—this time with the new word it did not understand—"recalcitrant." The idea behind the call to the function is that once it understands the new word, it can use that knowledge to understand the meaning of the first word, "obstreperous."

15.3 HOW RECURSION IS EXECUTED

Recursion is dependent on the ability of a program *to call a function over and over again*. The first step in recursion is to define a function that will solve *the first case* of our problem.

COPIES OF FUNCTIONS ARE GENERATED

Each time a call is made to a function, the compiler will generate *a copy* of the instructions (the code) of the function in the memory of the computer. If you call a function repeatedly, then *for each call* there will be a copy of the function stored in memory. Let's say a function called "Alpha" (with an integer parameter) is called three times during a program's execution.

```
int x, y, z;
. // x , y, and z are assigned before the calls are made
.
.
Alpha (x);
.
.
.
Alpha (y);
.
.
.
Alpha (z);
```

"Alpha's" code will be copied three times in memory as each copy is made for each call (Figure 15.3).

ONCE THE CALL IS MADE

Whenever a call to a function is made, the compiler immediately leaves the place where it is to go to the place that was called. (Recall that this *call* to a function just needs the *function name* with any *necessary parameters* inside.) In the main function of the program, there will be an *initial call* to

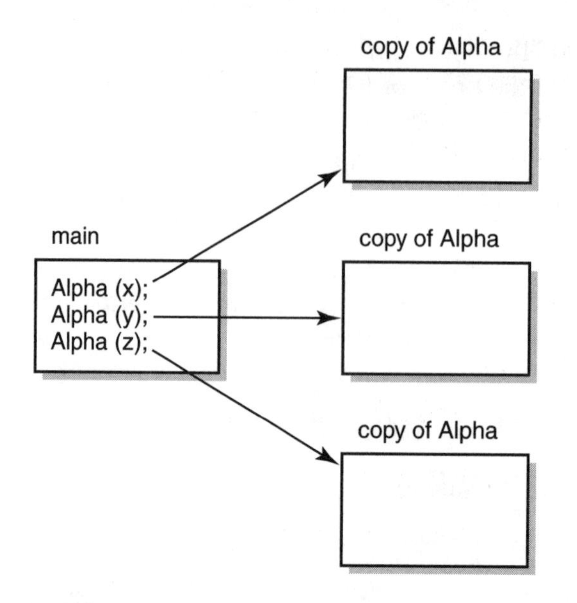

FIGURE 15.3 Three copies of the function "Alpha" are shown for each call made to the function.

the factorial function *perhaps followed by an output statement to show that answer on the screen.* The *initial* call is the *first* call to start the recursive process. After recursion is completed, the next statement—here, a "cout" statement—is executed.

```
result = factorial (6); // a function is called here
cout << result << endl; // the answer will be on the screen
```

Once inside the function "factorial," the recursive process begins because the function includes a call to itself. However, this time the call is slightly different because "factorial" is called with a *smaller number* as a parameter. The statement that contains the call will look something like this statement:

```
answer = 6 * factorial (5);
```

compiler leaves this statement to go to the factorial function

In this statement, the variable "answer" is assigned a *product* (the answer you get when you multiply two numbers). But the product, 6 * factorial (5), cannot be calculated because the computer does not know the answer to the factorial of 5! Before the product can be found, the compiler will leave the function it is in and go to another copy of the function to compute the value of the factorial of 5. We are doing the problem for a second time, but this time we are computing the factorial for a smaller number, 5 (Figure 15.4).

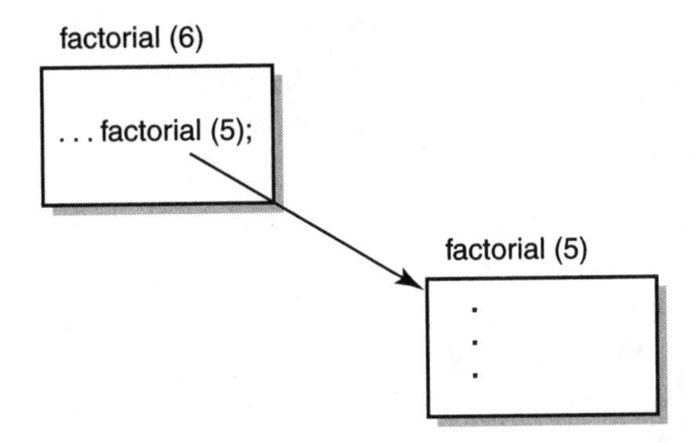

FIGURE 15.4 The function is shown with the value that was sent into it (6) and then another copy of the same function is shown with a new (smaller) value (5) that was sent into it.

When the second function finishes its work in computing the factorial of 5, *it will return a value to the exact place where it was called*. The returned value is used to compute the product in the assignment statement.

```
answer = 6 * factorial (5); // inside factorial "6"
         // 6 * some value will be assigned to answer
```

What happens in the "factorial" function that had 5 sent into it? That function (we'll call it factorial "5") had its own assignment statement with a *call* to another function. Consider this statement:

```
answer = 5 * factorial (4);// inside factorial "5"
```

It looks like the exact same statement we saw in the *first copy* of the function (the "factorial" function with the parameter, 6). Now we are in another copy of the function factorial "5." This statement includes another call to the factorial function—this time with the parameter 4. Let's call it factorial "4."

So the factorial "4" is called and we can expect to see the same kind of statement inside the function as we have already seen.

Here is the same assignment statement in factorial "4":

```
answer = 4 * factorial (3);// inside factorial "4"
```
the second last call in recursion

We are almost done with these calls. There are two calls left before we reach the "stopping state"— the end of recursion. The second-to-last call is the call to the factorial function with the parameter 3 sent into it. (Let's call this function copy, factorial "3.") Once you enter the factorial "3" function you will see *the last call* in recursion. This is the statement:

```
answer = 3 * factorial (2);// inside factorial "3"
```
the last call in recursion

Now the last call has been made and we enter the factorial with parameter 2. Here is the "stopping state" of recursion. We have encountered a factorial problem whose answer we do know; we do not have to call anything else to get an answer. We have a "partial" answer to our problem.

```
answer = 2 * 1;// inside factorial "2"
```
no more calls

COMING BACK UP THROUGH A RECURSIVE WEB

We have a partial answer to the original problem. We now need to examine the next stage of recursion to see how we arrive at the final (definitive) answer to the original problem, computing the factorial of 6 (6!).

So far we have just been *calling* other functions and not getting any results. With the call to factorial "2," we have an intermediate answer—the value, 2, from the product of 2 * 1. The way to get the final answer is to check what happens next in each function. Consider this fragment of code from factorial "2."

```
answer = 2 * 1;// inside factorial "2"
return answer;
```

What happens with the *return* statement? If you recall, it will cause the value in the variable "answer" to be sent back *to the place that called it*. Unlike all of our previous examples where we used the *return* statement, we were always sending values *back to the main function* (where the call was begun) of the program. In this example, we are sending the value back to the previous factorial function, factorial "3."

Watch what happens when we look at the statements in "factorial 3."

```
answer = 3 * factorial (2);// inside factorial "3"
                   ⇓
                   2

return answer; // from 3 * 2
         ⇓
         6
```

The value 2 is multiplied by the number 3 because 2 was *returned* from factorial "2." But wait! After this statement, there is another *return statement* that will cause another value to be returned to the place that called it. This time the value 6 (3 * 2) will be returned to the place that called factorial "3." But which function called factorial "3"? It was factorial "4" that called factorial "3" and here are two statements from factorial "4."

```
answer = 4 * factorial (3);// inside factorial "3"
                   ⇓
                   6
return answer; // from 4 * 6
         ⇓
         24
```

The value 6, which was computed and returned from factorial "3," is multiplied by the value 4. Next, the value 24 will be returned to the function that called factorial "4"—factorial "5." The same process is repeated inside of factorial "5."

```
answer = 5 * factorial (4);// inside factorial "5"
                   ⇓
                   24
return answer; // from 5 * 24
         ⇓
         120
```

We have almost finished following the compiler back to all the places where calls were initiated to other functions. Now the value 120 will be returned to the function that called factorial "5," which was factorial "6."

```
answer = 6 * factorial (5);// inside factorial "6"
                   ⇓
                   120
return answer; // 6 * 120
         ⇓
         720
```

The final statement in factorial "6" will return the value 720 to the main function, where the first function call was made. Once inside the main function, that value is printed on the screen through the "cout" statement.

```
result = factorial (6);
            ⇓
           720
cout << result << endl;  // 720 is printed on the screen
```

This is just one example of how recursion, the process of a function calling itself, is used to solve a problem. Recursive solutions tend to be very quickly computed but have the disadvantage of using up a lot of memory. Remember that every time a function is "called,", another copy of that same function (this time with a different parameter) is generated.

HINT!

Recursion always involves "calls" to the same function with different parameters each time. Another important part of recursion is understanding how the compiler moves from one function to another through the return statement that is used in each function.

ON THE CD-ROM!

A program (written in the C++ programming language) that computes the factorial of a number.

SUMMARY

This chapter covers the topic of *recursion*—a method of solving a problem by using a "smaller" case of the same problem. Functions are used in recursive solutions. Although the same function is called over and over again, it is called with a different parameter each time. The value of the parameter is *smaller* for each successive call and this is what is meant by a "smaller case" of the same type of problem. Eventually a "smallest case" is reached, called a "stopping state," and the recursive process ends.

After all the calls have been made, the *return* statement of each function sends the compiler from one function back to the function that originally called it. This "chain" of recursion is what makes the process sometimes difficult to follow.

We covered two examples in this chapter. The first example was a situation where the meaning of an unknown adjective is found by repeatedly looking up simpler adjectives of the original word. In the second example, a factorial operation on a number is done by finding the answers to factorial operations of smaller numbers. A factorial of a number is defined as a product of the number and all numbers less than it, down to the number 1.

CHAPTER

16

HTML: High Level Languages and Why They Are So Easy!

IN THIS CHAPTER

- Why Did HTML Come about?
- What Is It Used for?
- HTML: A High Level Language
- Definitions of Tags, Attributes, and Values
- Nested Tags
- Creating an HTML Document
- The Header and Title
- The Linking Tag: How to Bounce to Another Page
- The Image Tag: How to Insert a Photo

16.1 WHY THIS LANGUAGE NOW?

HTML is one of a group of languages that have arisen because of the widespread use of the Internet. Let's take a moment to explain a little about the motivation for the existence of this language and then we can talk about features of the language.

If you consider the Internet, it is a wide network of computers all around the world. It is actually a network of other networks. The biggest advantage of the Internet is that a programmer can send his files to another computer, called a remote computer. Ten or fifteen years ago, sending files between two computers always posed problems. There were concerns that the remote computer did not share the same applications as the computer on which the files were generated.

If you had any experience with this situation yourself, you might remember sending a file to a friend and then having that friend remark that all he received was some gibberish in the file. This was a frustrating, common experience (Figure 16.1).


```
V□æ□B□□□ç□B□□□□□□□□□
(A□dB□□CHDHEHFHHL$`□a□b□l˘˘˘˘˘˘□□□B□□□□□B□□□˘˘□□□□Geneva□□Ti
mes□□HH□/□(˘·˘,□˘□F□G□(□¸□HH□/□(□d□□□□H'□□□□□□ê@□□□□□˘□£□d
£□□˘˘˘ä˘˘˘'□□□□□˘˘cc}à□¶NÄ
  AÄ□x ImportPNGCanDoø¡˘¯l□¶;‰ê□î!˘∞8`□□H,
ÄA□;√H,□ÄA□lc□58û@ÇT8□H,□ÄA□<gr0`ipì¸ì¸□8ÉH,□ÄA□,□ê~8ü@Ç□8`
¯,ÅÅX0!PH□H,□ÄA□ÅÅX0!P}à□¶ª¡˘¯NÄ    AÄ□□ú
ImportPNGOpenì·˘¸;„l□¶,-
Ä□ê□,Éî!˘∞AÇ@AÜH+µÄA□Ä□□□H+¡ÄA□Ä□□□H+µÄA□Ä□□□H+©ÄA□8□H+µÄA□
8`ÅÅXÉ·L}à□¶0!PNÄ    AÄ□□xImportPNGClose<□0`      NÄ
@□□ImportPNGVersionêÉ□8`NÄ
@□□ImportPNGTargetøA˘Ë;§;c,ùMM□□□˘˘˘˘˘˘˘˘˘˘˘˘˘˘˘˘˘˘˘˘˘˘˘˘˘˘
˘˘˘˘˘˘˘˘˘˘˘˘˘˘˘˘˘˘˘˘˘˘˘
```


FIGURE 16.1 A typical transmitted file with some of the nonsensical symbols contained in it.

At that time, the only successful transmissions of files would be between two computers that had the same applications and operating systems and, therefore, could both "understand" the common files sent between them.

Another way of handling this problem was to send files as "text" files. From Chapter 11, you might recall our discussion of text files, where they contain only text (i.e., the characters found on a keyboard). But text files are limited because they present words or strings as just a long stream of letters. Text files contain no directions for the format of how that text should look. The advantage of the text file was that you could get your message across to the remote computer, but forget about having that file look good on the other end.

HTML is the answer to this problem. It is the first of the "Internet-driven languages." This language and others like it (SGML, XML, and Java to name a few) owe their design and existence because they were created to solve the transmission of all kinds of information over the Internet among varying types of computers.

How can we send text files so that they look good when they get to the other computer? Hypertext Markup Language, or HTML, provides the answer to this problem.

A REALLY HIGH LEVEL LANGUAGE

It has been a while since we have spoken of high level languages—not since Chapter 1. HTML (the acronym for Hypertext Markup Language) is an example of a very *high level* language. Contrary to what you might think, this language is actually easier than the other languages of C, C++, and Java, which you will no doubt study at some point. The adjective "high level" refers to the fact that the programmer has actually less to do than the programmer of a low level language. High level programming is really "hands-off" programming.

THE ABSENCE OF TYPICAL HALLMARKS OF PROGRAMMING LANGUAGES

HTML has no loops, no control statements such as the if...statement or the if...else statement we studied previously. The most that a programmer will do is assign values to certain variables so that certain things can occur. For this reason, a learner of this language has little of the work essential in learning most other programming languages.

16.2 TAGS, ATTRIBUTES, AND VALUES: FEATURES OF HTML

To understand the main properties of HTML, we need to examine three key parts of this language: tags, attributes, and values. A tag (just like its meaning in everyday English) is something you attach to something else to give it some extra meaning or marking.

Think of all the price tags you see in the stores every day. Each tag has all kinds of information—an identification number, a price, a classification of item type, and a size are just a few of the items we can see on a price tag.

In HTML, tags are used to *mark* text (i.e., they are used to give some significance to what would otherwise be a stream of letters or words). Let's consider some of the tags that exist in HTML.

FUNCTIONS OF TAGS: EXAMPLES OF TAGS

There are tags that cause a new paragraph to be placed at a certain point in a block of text. Another tag, as an example, is used to insert a graphics image. There are tags that change the font style and size of the text so that you can see bigger or smaller letters. Another tag causes a line feed, in case you wish to have a sentence be displayed on more than one line. (A line feed causes the cursor to move to the next line.)

Thus tags are used to format the text (i.e., give the text some structure). Normally when a computer "reads" a text file, it does not do anything more than look at the information in the file (usually a code number called an ASCII code) and then translate that code (the ASCII code) into the appropriate symbol from the keyboard (or from the ASCII list).

HOW TAGS WORK

A tag is represented with two parts: an opening tag and a closing tag. An "opening tag" will appear at the beginning of the text it is affecting and then a "closing tag" appears at the end of the text affected by the tag. A tag is comprised of a letter inside of a set of angle brackets (< >): <B>,</B>, <I>, </I>, <A>, </A>. The forward slash in a tag within the brackets is used to represent the "closing tag."

Let's use the paragraph tag as our first example. The capital letter "P" is the name of this tag. The tag will only be used at the beginning of the text it is supposed to format. It causes a line feed so subsequent text appears on the next line. (The closing tag for the paragraph, </P>, is unnecessary.) Here is an example of text written in HTML and using the paragraph tag.

EXAMPLE OF HTML TEXT

```
<P> Once upon a time there was a greedy king. He tried to gather
all the jewels in the country to keep in his castle. He even took
```

```
the jewels from the queen and kept them in his safe. <P> In one
town in the kingdom, a young woman had a pendant that was given
to her by her godmother. She did not want the king to know that
such a pendant existed.
```

How the text will appear when viewed on a remote computer:

Once upon a time there was a greedy king. He tried to gather all the jewels in the country to keep in his castle. He even took the jewels from the queen and kept them in his safe.

In one town in the kingdom, a young woman had a pendant that was given to her by her godmother. She did not want the king to know that such a pendant existed.

The paragraph tag is just one of several tags in the language. Using the tag is a way of giving instructions to another machine for how to structure the text. For this reason, HTML commands need to be universally understood by the computers that send and receive these files.

Another example of the tag is the italics tag, represented by "I." When the opening and closing tags for italics are used around a block of text, all text within the block is *italicized*. The "bold" tag, represented by 'B,' will cause text within its opening and closing tags to be boldfaced. Consider this example, which uses both tags.

EXAMPLE IN HTML TEXT

```
<P> <I >Once upon a time there was a greedy king. He tried to
gather all the jewels in the country to keep in his castle. He
even took the jewels from the queen and kept them in his safe.
</I> <P> In one town in the kingdom, a young woman had a pendant
that was given to her by her godmother.<B> She did not want the
king to know that such a pendant existed.</B>
```

As the text appears on a remote computer:

Once upon a time there was a greedy king. He tried to gather all the jewels in the country to keep in his castle. He even took the jewels from the queen and kept them in his safe.

In one town in the kingdom, a young woman had a pendant that was given to her by her godmother. **She did not want the king to know that such a pendant existed.**

ATTRIBUTES

Attributes are defined as additional qualities within a tag's function that add more information to the tag's operations. Attributes are then assigned values. Consider the *font* tag. The *font* tag encompasses all aspects of the writing style of text. There are a few things to consider about writing style, however. In what type *face* should the text be written? How big do we want the text to appear? This is the *size* of the font. In what color do we want the text to appear? A short list of a tag and its attributes follows:

Tag	*Attributes*
Font	Face, Size, Color

Face, size, and color are all additional aspects of the general font tag. To use a font attribute, we list it and give it a value within the tag itself (inside the angle brackets). Consider this example.

```
<FONT   FACE = "Courier" >                         opening font tag
  tag   attribute assigned a value

</FONT>                                            closing font tag
```

When the FACE attribute is given, the font style becomes the style of the value that is given in quotation marks. (The attribute is like a variable name that is assigned a value.) Let's apply the expanded tag to our previous example about the "greedy king."

EXAMPLE IN HTML TEXT

```
<P> <FONT  FACE = "Courier" >  Once upon a time there was a
greedy king. He tried to gather all the jewels in the country to
keep in his castle. He even took the jewels from the queen and
kept them in his safe. <P> In one town in the kingdom, a young
woman had a pendant that was given to her by her godmother. She
did not want the king to know that such a pendant existed.
</FONT>
```

How this appears on a remote computer:

```
Once upon a time there was a greedy king. He tried to gather all
the jewels in the country to keep in his castle. He even took the
jewels from the queen and kept them in his safe.
```

```
In one town in the kingdom, a young woman had a pendant that was
given to her by her godmother. She did not want the king to know
that such a pendant existed.
```

Let's use another example with two font attributes: the color attribute and the size attribute. The color attribute will affect the color in which the text is displayed. Color values are the names of the colors themselves or the hexadecimal numbers (code numbers written in base 16) used to represent those colors. The size ranges from one to seven. (For your information, most of this book appears in "Times" style.)

EXAMPLE IN HTML TEXT

```
<P> <FONT  FACE = "Courier" SIZE =  "3" COLOR = "red">  Once
upon a time there was a greedy king. He tried to gather all the
jewels in the country to keep in his castle. He even took the
jewels from the queen and kept them in his safe. <P> In one town
in the kingdom, a young woman had a pendant that was given to her
by her godmother. She did not want the king to know that such a
pendant existed. </FONT>
```

How that text appears on a remote computer:

```
Once upon a time there was a greedy king. He tried to gather all
the jewels in the country to keep in his castle. He even took the
jewels from the queen and kept them in his safe.

In one town in the kingdom, a young woman had a pendant that was
given to her by her godmother. She did not want the king to know
that such a pendant existed.
```

Nested Tags

A "nested" tag is a tag that is applied to some text that is already being worked on by another tag. When you start to use another tag *within an existing tag's text block*, the only thing you need to remember is to close each of the tags when you are done applying them. It's possible for one tag to keep operating while another has been closed. We'll use a centering tag to center the text in the middle of the page and at the same time, we'll insert paragraph tags and a boldface tag. As an example, let's return (once again!) to the short (very short!) story about the greedy king.

EXAMPLE IN HTML TEXT

```
<P> <CENTER> Once upon a time there was a greedy king. He tried
to gather all the jewels in the country to keep in his castle. He
even took the jewels from the queen and kept them in his safe.
<P> In one town in the kingdom, a young woman had a pendant that
was given to her by her godmother. She did not want the king to
know that such a <B>pendant </B> existed. </CENTER>
```

How that text appears on a remote computer:

Once upon a time there was a greedy king. He tried to gather all the jewels in the country to keep in his castle. He even took the jewels from the queen and kept them in his safe.

In one town in the kingdom, a young woman had a pendant that was given to her by her godmother. She did not want the king to know that such a **pendant** existed.

It's even possible to apply tags *within one word.* Let's take the short phrase "greedy king" and change the color and the size within the word itself.

EXAMPLE IN HTML TEXT

```
<P> <CENTER> Once upon a time there was a gr<FONT SIZE = "6"
COLOR = "red">ee</FONT>dy king. He tried to gather all the jewels
in the country to keep in his castle. He even took the jewels
from the queen and kept them in his safe. <P> In one town in the
kingdom, a young woman had a pendant that was given to her by her
godmother. She did not want the king to know that such a
<B>pendant </B> existed. </CENTER>
```

How that text appears on a remote computer:

Once upon a time there was a greedy king. He tried to gather all the jewels in the country to keep in his castle. He even took the jewels from the queen and kept them in his safe.

In one town in the kingdom, a young woman had a pendant that was given to her by her godmother. She did not want the king to know that such a **pendant** existed.

HINT!

Remember that the paragraph tag has an *optional* closing tag.

DEPRECIATED TAGS

The term "depreciate" usually means to go down in value. Stocks can depreciate as can the value of a new car over time. The usefulness of the most "up to the minute" computer will depreciate over time because newer, faster computers are always being developed.

Depreciated tags in the HTML language are tags that are no longer as useful as they once were. Newer versions of "browser" applications will not necessarily understand the depreciated tag. For this reason, it is important that, as you begin to learn HTML, you check the most recent standards of the language. (A Web site dedicated to dissemination of standard HTML is the following: http://www.w3c.org.)

16.3 CREATING AN HTML DOCUMENT

So far in our discussion of tags, we have focused on those used once a document has been begun. Now it's time to create an HTML document by focusing on the tags necessary to do this. The first tag to use is the HTML tag. Both the opening and closing HTML tags are used to surround an entire HTML document. Any document will show these two tags at the beginning and end of the document.

```
<HTML>    opening tag



</HTML>   closing tag
```

Because the main purpose of an HTML document is to create a Web page that someone else will be viewing on his or her computer, we need certain other components for these pages.

HEADER, TITLE, AND BODY TAGS

Within the two HTML tags go the entire text document that we are structuring. The next tag used is the *header* tag (<HEAD>, </HEAD>), used after the HTML tag for the sake of the browser—the application responsible for bringing our Web page to the remote computer. The header tag includes the title of the Web page.

The title tags are used inside the header tag to specify the title of the Web page we are developing. Web pages should have a title. In that way, anyone who visits our Web page will know what the topic or main points of our page are. Furthermore, when a search is performed by a search engine, the engine will be able to scan the information within our header tags to see if we are a page that fits the constraints of the search.

Imagine if we wrote a Web page about the history of kings and why they are often greedy—then the title of our page could be "The Apparent Greed of Kings in Modern Day Ehrichia." The title tags are used for the benefit of a search engine of our page. The title tags are as follows: <TITLE> and </TITLE>.

If you want to insert a heading at the top of your page, you need to use another header tag. This time you number the header and the tag looks like this: <H1>, </H1>. The number "1" indicates that this is the largest header on the page itself. Compare the placement of the title of the page and a header within the document itself (Figure 16.2).

The BODY tags are used to surround the main text of the document, referred to as the body of the document. Just like the previous tags, the

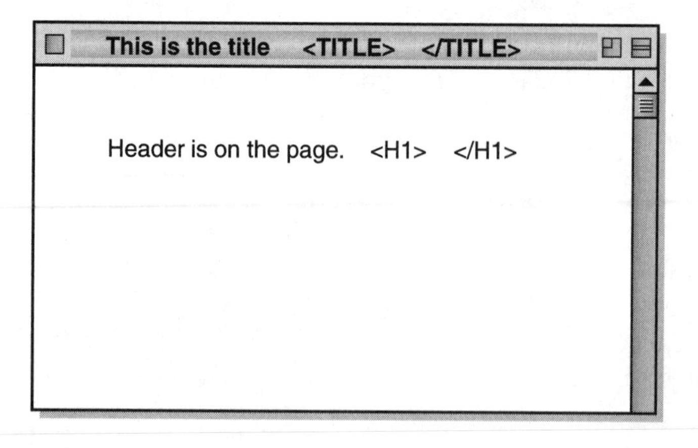

FIGURE 16.2 A title appears at the top of the bar whereas a header appears within the document itself.

opening tag is used to begin the body of text and the closing tag is used after the text (<BODY> and </BODY>). The BODY tag has an important attribute, called the LINK attribute, which we will examine later. Look at the following list of tags and the comments on the side to clarify which tag is which.

```
<HTML>     //HTML document begins here
<HEAD>     // Head of document begins here
<TITLE>    The Apparent Greed of Kings in Modern Day Ehrichia
</TITLE>   // title begins and ends on one line usually
</HEAD>    // close of the head of the document
<H1>       The Once and Greedy King </H1>
<BODY>     // the beginning of the main part of the document
              called the body
           // the body of text goes here
</BODY>    // the closing of the body of text
</HTML>    // the end of the HTML document
```

We can now complete the HTML document by merging the tags used in this example with the body of text presented previously to create an entire document.

EXAMPLE OF HTML FILE

```
<HTML>
<HEAD>
<TITLE> The Apparent Greed of Kings in Modern Day Ehrichia
</TITLE>
</HEAD>
<H1> The Once and Greedy King </H1>
<BODY>
<P> <FONT  FACE = "Courier" SIZE = "3" >  Once upon a time there
was a greedy king. He tried to gather all the jewels in the country to
keep in his castle. He even took the jewels from the queen and kept
them in his safe. <P> In one town in the kingdom, a young woman had
a pendant that was given to her by her godmother. She did not want the
king to know that such a pendant existed. </FONT>
</BODY>
</HTML>
```

How that text appears on a remote computer:

```
 The Once and Greedy King

Once upon a time there was a greedy king. He tried to gather all the
jewels in the country to keep in his castle. He even took the jewels
from the queen and kept them in his safe.

In one town in the kingdom, a young woman had a pendant that was
given to her by her godmother. She did not want the king to know that
such a pendant existed.
```

EXERCISES

Insert tags around each of the following phrases so that each action will occur.

1. Boldface "My Best Friend."
2. Italicize the adverb in "He appeared surreptitiously!"
3. Create a title for a Web page about "The Worst Rock Groups of the 1970s."
4. Change the size of the following phrase to size 16: "Last Summer in Brisbane."
5. Change the font style to Helvetica for "Apple Picking in Massachusetts."

16.4 THE INTERNET, BROWSERS, AND URLS

HyperText is the ability of text to be able to "bounce" from one file to another on the Internet. The text has the power to actively "move" the user to another location, which is usually another web page. This is done through the linking tag, which allows another site to be introduced.

To understand how hypertext works, we need to examine some other commonly used terms with which you may or may not be familiar. These terms refer to the different structures that exist so that computers can allow this ability to "move" from one location to another.

BROWSERS

Browsers represent a particular kind of application program. (Recall that applications are programs that are used to do something useful and very specific such as a word processing application, a game-playing application, and balancing checking accounts). Browsers are designed to retrieve files from one computer and bring them into the user's computer.

Most of you are familiar with Netscape *Navigator*™ and Microsoft *Internet Explorer*™. These are two Web browsers that are able to respond to commands to get files and retrieve them over the Internet. Obviously, the Internet is more than a network of computers. In a sense, it is a network of many other networks—which is why it is such a massive grouping of computers.

So, when we write a document in HTML, we are really writing it for the sake of the browser so that it can both understand our document and be able to access it for someone else who might wish to visit it on the Internet. When you visit a Web page that someone else writes, you can get there two different ways. You might arrive at a location through a search engine or via a link from another page.

To be an effective language, HTML must be standardized so that all computers with browsers can understand its commands.

URLS

A URL is an acronym for Universal Resource Locator. The Universal Resource Locator includes the address of a specific Web page and also specifies other information for the browser. Let's examine the three parts of a URL.

PROTOCOL

When a browser goes to get another file, it needs to know the protocol by which it obtains the file. *Protocol* in English usage is the way we do something or another word for protocol is procedure. What is the *protocol* for ordering tickets to a concert over the phone? The protocol is to call the 1-800 number with your credit card number in hand and your list of desired dates. What is the protocol for dealing with this file's address? I want to use the right protocol for the right computer so that the other computer "understands" me.

Here are some examples of the most common protocols used by browsers:

ftp ://	ftp stands for file transfer protocol
http://	stands for hypertext transfer protocol
gopher://	stands for "gopher" protocol

The common *procedure* or *protocol* for all computers that are linked on the Internet is TCP/IP or Transmission Control Protocol/Internet Protocol. In a sense, this is the way computers on the internet "talk" to each other. They speak "TCP/IP."

HOST/SERVER NAME

The second part of a URL is the host name (server name) of the computer where the file you want is located. Most host names begin with the common World Wide Web prefix: for example www.location.com or www.location.org. This name is called the domain name system or DNS for short. The last word (e.g., .com, .org, .edu, etc.) refers to the zone of the organization—a commercial, military, or educational zone, for example.

There are computers at universities and schools; these would most likely have DNS's that end in ".edu." Computers at companies or commercial organizations use the ".com," which most of us have heard as "dot com."

RESOURCE

The third part of the URL is the file or resource you wish to access from the host computer. Frequently this is a file name accessed through a directory of folders that contain it. Here is an example of an imaginary URL.

```
http:// www.mycompany.com/products/books/topics.htm
```
transfer protocol server nam folder name another folder name a file name

The browser now knows everything it needs to know about how to locate the particular file (an html file with the extension ".htm"). The file "topics.htm" is found in the "books" folder, which is found within the "products" folder, which is found on the host computer (a commercial group denoted by ".com"). The commercial group's name is "my_company" and it is found on the World Wide Web. The file will be transferred according to "hypertext transfer protocol" specifications.

Here are two other examples using other protocols besides the HTTP. These are two different (again, imaginary) URLs:

ftp:// ftp3.mysecondcompany.com/cplusplus/firstprog.cp
gopher://gopher.myuniversity.edu/whatisthere.lis

These other protocols are used to gain access to computers that might not be part of the World Wide Web listing (hard to imagine). The gopher protocol is an older protocol that some computers still use.

It is important to remember that the word "Internet" refers to interconnected networks of computers. These networks include computers that respond to "gopher" protocol or "gopher" behavior, and computers that understand "file transfer" protocol. In recent years the World Wide Web, which understands "hypertext transfer protocol" and Web pages written in HTML, has become the biggest network of computers on the Internet and the most widely used. As a result, most people assume it is the only network of computers on the internet and that HTTP is the only protocol that computers understand.

IMAGE RESOURCES

We can use a URL to access any resource on the Web, including a graphics file. All we need to know is what the graphics file is called and where it is located. Then a browser can access the file. Recall that graphics files usually have the extension ".gif," ".jpeg," or ".jpg."

EXAMPLE

```
http://www.greatnetwork.net/~myname/mypics/greatphoto.gif
```

16.5 ACTIVATING HYPERTEXT THROUGH A TAG

An important tag in the HTML language is a linking tag that activates the HyperText properties of the language (text that activates the opening of another window or the transportation to another file on the Web through "clicking," for example).

The link tag is the letter 'A' and is enclosed within the angle brackets like the other tags we have examined—<A> and </A>. The *attribute* of the A tag that allows us to move to another Web page is the HREF *attribute*. We can expand the tag by using this *attribute* and giving it a *value* as we do all attributes.

```
<A HREF = "  URL goes here ">  LABEL</A>
```

opening tag   label   closing tag

```
<A HREF = "http://www.myCompany.com ">  My Company's Home Page </A>
```

opening tag   URL   label   closing tag

LABELS

In between the opening and closing tags that designate another location to visit, we insert a *label* in our HTML document. The *label* is just one of our text words (from our document) which, when clicked, will activate the browser to go to this other page. This *label* will be what the user sees on the page. The label should be a word that relates to the page we want to visit.

The label is always *highlighted* in a different color from the main text so the user will notice it. We can designate the color for the label by using an attribute inside the BODY tag of the HTML document. Recall that the *BODY* tags surround the text that comprises the main part of our Web page. The body follows the *head* tag of the document.

The attribute is called the LINK attribute and it is given a color value within quotation marks. Here is an example:

```
<BODY                LINK = "blue">
opening body tag  link attribute  color of link
text follows

</BODY>
```

HINT!

The user doesn't see the address of the Web page but only the label, which, upon clicking, sends him to the designated address.

So let's say in our document about the greedy king, we wish to refer our readers to another Web page that has information about pendants. (We are anticipating that a reader might not know what a "pendant" is, so we want to link to another page that has that information for the user.) We will insert a URL for pendants into the document and it will look like the following example.

EXAMPLE IN HTML TEXT

<P> <FONT FACE = "Courier" > Once upon a time there was a greedy king. He tried to gather all the jewels in the country to keep in his castle. He even took the jewels from the queen and kept them in his safe. <P> In

one town in the kingdom, a young woman had a <A HREF = http://www.jewelry.org/necklaces/pendant.htm>pendant</A> that was given to her by her godmother. She did not want the king to know that such a pendant existed. </FONT>

```
Once upon a time there was a greedy king. He tried to gather all the
jewels in the country to keep in his castle. He even took the jewels
from the queen and kept them in his safe.

In one town in the kingdom, a young woman had a pendant that was
given to her by her godmother. She did not want the king to know that
such a pendant existed.
```

IMPORTING IMAGES

Another important feature of HTML is the ability to have items other than text in our document. Web pages become more creative and interesting each day. Let's say we want to be able to put a photo into our Web page. Here we need to use the image tag, which will allow us to access an image we have saved on our computer.

The *image* tag has an attribute, called a *source*, where you give the name of the graphics file. The source will have a name followed by an extension. Most images will have a ".jpeg," ".jpg," or ".gif" extension ("jpeg" stands for "joint photographic experts group" and "gif" stands for "graphical interchange format").

Each of these extensions is a type of graphics file and each stands for a different way of storing data to represent an image.

When you incorporate an image into a document, you need to consider that another computer viewing your Web page might not have the capability to read this file. So we insert an alternate text sentence to be inserted if the graphics file cannot be read. The text is put in quotation marks and it is assigned to the attribute, ALT. Here is an example:

```
<IMG SRC = "greatphoto.gif" ALT = "Picture of a very greedy king">
```

tag attribute filename alternate tag with text

If the file "greatphoto.gif" could not be read by another computer then the alternate text "Picture of a very greedy king" would appear in its place.

When a graphics file is inserted into an HTML document, we need to decide how big that photo should be—whether it should take up the whole page or only part of it. To set the size of the picture, you need to assign values to two attributes within the image tag: the width and height attributes.

Both the width and height attributes must be given values that correspond to the number of pixels on the screen. If your screen is 1024 pixels wide by 768 pixels high, you can estimate the width and height of your image from those measurements (Figure 16.3).

A Picture Within an HTML Document

Compare the size of this image with the entire page.

This figure is 112 pixels wide and 74 pixels high.

FIGURE 16.3 An image is shown with its width and height given in pixels in relation to the screen that is 1024 × 768 pixels.

ALIGNING TEXT

Before inserting the image into the document, we need to consider one additional aspect. Where should the photo be placed? We can align the photo, that is place it to the right side of the page or to the left side of the page by using the align attribute within the image tag. Then we give the "align" attribute a value of *right* or *left*. Then the photo will be put on either the left side of the screen or the right side of the screen, according to the value chosen.

Let's consider an example that uses both the height and width attributes and the align attribute. Consider the following HTML sample.

EXAMPLE IN HTML TEXT

```
<P> <FONT  FACE = "Courier" >  Once upon a time there was a
greedy king. <IMG  SRC = "greatphoto.gif"  ALT = "Picture of a
very greedy king"  WIDTH = "100" HEIGHT = "200"  ALIGN = "left">
He tried to gather all the jewels in the country to keep in his
castle. He even took the jewels from the queen and kept them in
```

```
his safe. <P> In one t own in the kingdom, a young woman had a <A
HREF   http://www.jewelry.org/necklaces/pendant.htm> pendant </A>
that was given to her by her godmother. She did not want the king
to know that such a pendant existed. </FONT>
```

```
Once upon a time there was a greedy king. He tried to gather all
the jewels in the country to keep in his castle. He even took the
jewels from the queen and kept them in his safe.

In one town in the kingdom, a young woman had a pendant that was
given to her by her godmother. She did not want the king to know
that such a pendant existed.
```

16.6 HOW TO GET AROUND DOING EVERYTHING YOURSELF

In reality, there are several applications that facilitate the creation of HTML documents. You no longer have to type everything yourself. One such application is "Adobe *Page Mill*" or any HTML editor. These applications are designed so that you can easily construct your own page using a template or model that can be completed by you.

Even if you use an application that helps you create an HTML document, it is important to understand the structure of the HTML document because it will give you an appreciation of how Web pages are constructed.

ON THE CD-ROM

Some sample HTML files.

SUMMARY

We introduced the HTML language, which facilitates the transmission of text over the Internet. HTML allows text to be structured through the use of tags, attributes, and values. A tag acts like a marker placed on the text to cause some change in some aspect of the text. There are tags to cause paragraph breaks and tags that affect the style of the text. Attributes are additional qualities of the tag function and are inserted within the tag. Attributes are then given values.

An HTML document has a header tag within which the title tag is used. Then the body of the document follows. An important feature of the language is the ability to use hypertext, or text that allows us to move from one document to another. This is done through a linking tag. Another great feature allows us to insert photos (graphics files) into our document. We can also specify the size of the file and where it should be aligned.

Browsers are applications programs that are able to access files and transport them over the Internet. A browser works though a certain *protocol,* as specified in a URL or *Universal Resource Locator*. The information in the URL allows the browser to be able to "get" the file from another computer and transport it to our computer.

No doubt programmers will rely on applications programs such as Adobe *Page Mill* and other HTML editors to facilitate the development of an HTML document. It is helpful, however, to have an appreciation of the syntax of this high level language.

ANSWERS TO ODD-NUMBERED EXERCISES

1. <B>My Best Friend </B>
3. <TITLE>The Worst Rock Groups of the 1970's </TITLE> or <H1> The Worst Rock Groups of the 1970's </H1>
5. <FONT FACE ="Helvetica" >Apple Picking in Massachusetts </FONT>

CHAPTER

17

C++: Objects First, Classes Later!

IN THIS CHAPTER

- What Does C++ Offer as a Language—an Overview
- We Introduce the Object
- Define Object-Oriented Programming
- Define a Class
- Define Header, Implementation, and Client Files
- The Include Statement, Again
- The Public and Private Sections of a Header File
- The Object's Attributes and Some Examples
- Define Constructors
- Instantiating Objects through Calls to Constructors
- Using a Parameter List with a Constructor
- The Scope Resolution Operator

17.1 AN INTRODUCTION TO C++

The programming language C++, like many other languages, has all the control statements and data structures that you would expect a language to have. Control statements are those statements that allow the compiler to alter its normal line-by-line execution of statements. Loops are control statements and if... and if...else statements are control statements. Data structures are holders for data. Most programming languages offer integer, character, and real number holders for data as well as the struct type (also known as a record holder) we discussed in Chapter 10.

So what distinguishes C++ from all the other languages that exist now? (The first thing is that C++ is a well-established language that has existed since the mid 1980s, built upon the existing C language.) It is still a popular and well-used language so it is worth learning if you intend to be a serious programmer. As I have mentioned before, we need to see why a language comes into existence in order to appreciate its reason for existence. What we saw with HTML was that it was designed to solve a real problem of transmission of text (messages and documents) over the Internet. HTML was a solution that arose as a result of the wider use of the Internet.

The programming language C++ was designed to augment the capabilities of an existing language called C. C++ includes language capabilities to handle objects. *Objects* are data holders that have certain functions associated with them. In the next section, we look at objects in detail.

17.2 OBJECTS: WHAT ARE THEY AND WHAT DOES IT MEAN TO BE OBJECT ORIENTED?

There are different designs to programming languages depending on the way the language is used to solve problems. A language is designed so that a programmer can solve a problem within the confines of the language.

Languages that offer function design and syntax give rise to a programmer's solution to a problem. The programmer will think of his problem in terms of what functions he should write to accomplish different tasks. Function calls from one part of a program to another have also influenced the way programs are written. Programs become easier to read when you can focus on the big picture, which is visible through calls. Consider this section of a main program where these functions are called.

```
.
.
.
Load_arrayNames ;
Print_Names ;
Alphabetize_Names ;
Print_Names ;
.
.
.
```

Without knowing anything of the details of each of these functions, you can begin to guess what happens in each function call. The first function, "Load_arrayNames," probably loads values into the array. The next function, "Print_Names," prints those names. The third function alphabetizes the names that are in the array and then the print function prints those names again, this time in alphabetical order.

Function usage generally simplifies the code in the main function and makes that section easy to follow. All the details are *hidden away* in these functions. The design of a program that uses functions, alone, will focus on what each function should do—it is a *task-oriented* program.

Object-oriented design is a departure from this way of solving problems. To understand this design, we must first look at the definition of an *object*.

THE OBJECT AND OBJECT-ORIENTED DESIGN

The simplest way to define an *object* is to think of it as simply a *data structure*, or another *holder* for different kinds of data. In some ways, it is like the struct or record definition we examined earlier in Chapter 10. But an object signifies more than an ordinary variable. If we design a solution to a problem with the object as the focus of our design to a solution, then we say that the program is *object oriented*.

All functions in an *object-oriented* program are designed with the object as the "center of attention." How does a function serve the object? Rather than focusing on what the function *can accomplish as a task*, the focus turns to how the function *serves the object*. Furthermore, *all calls to functions are made through the object*. A program that is *object oriented* is analogous to a self-centered person who, always preoccupied with himself, can only think of everything else in the world in relation to himself.

Let's consider an example. If we take a problem of designing an address book, we would initially think of *actions* to take on the address book—like writing names and addresses into the book, updating an address, or changing a phone number.

What if we were to change the focus from the *actions* we take on the address book to the *address book itself*? What would an address book want to have done to itself? It would become the focus of our attention. (It would become the *object* in our *object-oriented* design.) Let's consider what the new focus of the problem would be. If the address book becomes the focus of design, let's consider how that changes the way we look at the problem. What would we do with an address book?

Things to do with an address book

Initially, create an address book.

1. Fill the address book with a new entry (a name, an address, and a phone number).
2. Alphabetize the address book-all the entries/slots get sorted.
3. Print the address book.
4. Count the number of entries in the address book.
5. Search for a name in the address book.
6. Search for a phone number in the address book.
7. Delete an entry in the address book.

From an *object's point of view*, the address book (as the object) calls on the different functions associated with it to keep the address book up to date and useful. Here is how an object-oriented design might look in a schematic drawing (Figure 17.1).

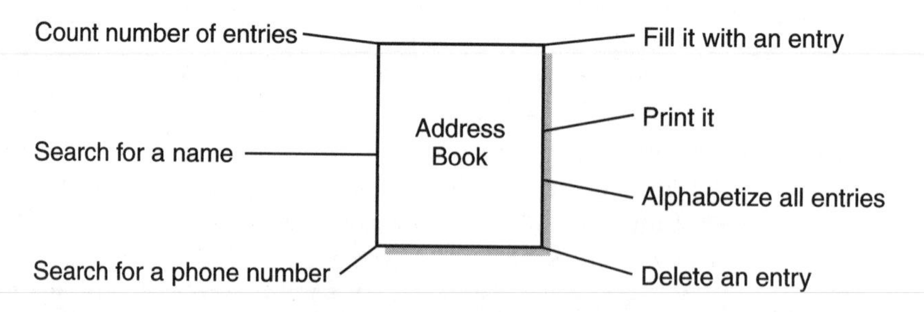

FIGURE 17.1 The address book is at the center of the problem and every other action is viewed in terms of how it affects the object, the address book.

HOW OBJECTS CALL FUNCTIONS

When an object calls a function, we say that the object is *invoking* the function, that is, it is initiating a call to the function. Because the object did the calling, it is assumed that the *function will work on that object* using any parameters (if any) in the call. Because the object calls the function—it remains the object of the focus of the attention. The call is done in the same manner as calling a regular function except that the object's name is used to *invoke* the function. Here is a model of how an object calls a function:

```
Object_name . function_name (any parameters in here);
```

The period is used to separate the object name from the function name. Inside the parentheses, which are always used in the call, parameters other than the object are listed. Note that even when no parameters are listed, parentheses must still be used. In the next section, we will examine the syntax of an object and its methods and where they are listed.

The object and each function associated with it belong to a larger entity called a *class,* which we examine in the next section.

17.3 CLASSES: WHERE IT'S AT

A class is defined as a grouping of an object and its functions. These functions are called member functions because they are members of the class and belong to it and are defined within it. All the information about the object data type and these functions is contained in the class. The class is defined and carried out through two separate files that contain all the code to handle the object and its methods. The first file is called a *header* file and the second file is called an *implementation* file.

The *header* file is where an object's definition as well as the headings for each of its functions are stated. The purpose of the header file is to give a reader a quick overview of all the member functions of the class and the design of the object associated with the class (Figure 17.2).

The *implementation* file gives much more detail for the reader. In this file, each function is shown with all its code in detail so that a reader can know exactly how the class is *implemented* or "carried out." The reason for separating the code into two files is to keep the class organized and easy to follow. The header file allows anyone to get a quick "overview" of the class without getting bogged down in the details, whereas the implementation file shows all the code (here, the programmer can see how the class is executed).

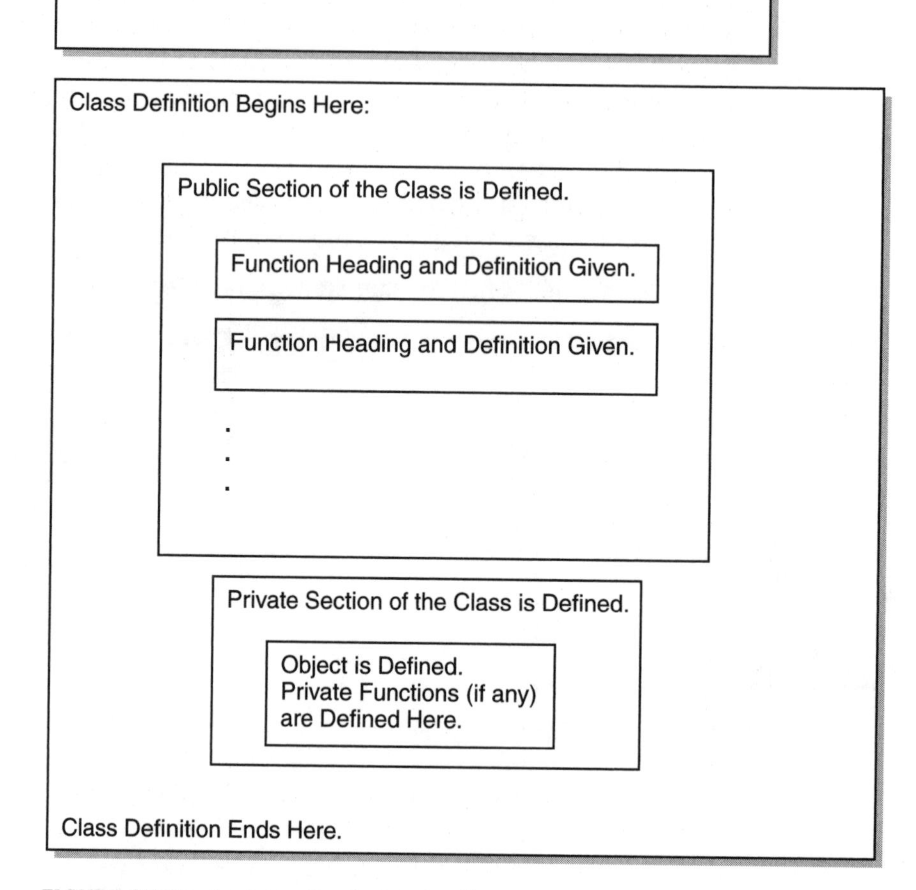

FIGURE 17.2 A schematic of a header file is given. Each member function's heading is given with a brief description of what that function accomplishes.

A third file associated with classes is called the *client file*. This file *uses* the class by declaring objects, which *access* the functions of the class. This file is a "client" of the class because it needs the class to properly execute.

HOW THE HEADER, IMPLEMENTATION, AND CLIENT FILES WORK TOGETHER

The client file is supposed to be a user-friendly file because it is more readable; it does not contain all the code in the other files (the implementation file and header file). Now you might be asking why all the fuss about taking code that used to be in one program and splitting it among these three files. One reason is that code becomes easier to debug when it is compartmentalized as it is through a class. Another reason has to do with the fact that programs always need to be updated and object-oriented design is considered the least disruptive in that modification process.

Consider this example from a client file of an *Address_book* class. Without getting into too much detail regarding syntax, let's look at what that file might contain. As we mentioned before, the object will do the calling.

```
.
.
.
// my_book is the object
string newnum;

my_book.Add_entry ( );
my_book.Print_book( );
my_book.Change_Phone_Num (newnum);
.
.
.
```

It appears that the object, called *my_book,* is calling three of its member functions. The first one, *Add_entry,* probably adds another person's name, address, and phone number to the object. The second function, *Print_book,* prints all the names, addresses, and phone numbers in the *Address_book.* Then the third function changes some phone number (whose?) in the *Address_book* to the "newnum" variable used in the call.

This client file keeps all the code hidden from the user. This is why the program is user friendly. The user should not have any problem guessing what the client file is trying to do because the commands and calls seem clear. The code to carry out these operations, however, is hidden in the other two files.

HOW TO ACCESS THE HEADER AND IMPLEMENTATION FILES FROM THE CLIENT PROGRAM

Both the header file and the implementation file must be made accessible to the client file. This is done through the use of a compiler directive, the "include" statement, which we saw in Chapter 4.

The include statement is a directive to *get* a file, usually a header file, and *paste* that code at the top of the file we are writing. The extension, ".h," indicates that the file is a header file. In our previous example, an "include" statement should be used at the top of the client file. It would look like this:

```
#include "Address_book.h"
```

Another include statement should be used if we expect to get any input from the user or direct any output to the screen. Here is the second include statement that should be used:

```
#include <iostream.h>
```

So "iostream.h" is the name of a *class* header file. Now our previous program will use the include statement.

```
#include "Address_book.h"
#include <iostream.h>

int main ( )
{.
.
.
// my_book is the object
string newnum;
.
.
.

my_book.Add_entry ( );
my_book.Print_book( );
my_book.Change_Phone_Num ( newnum);
.
.
.
return 0;
}
```

An implementation file uses the same extension as other C++ files—".cpp," ("C plus plus") to distinguish it from the header file. The implementation file will be included *inside of the header file* so you do not need to include it at the top of the client file. By including these files in the client program, the compiler understands everything it needs to know about the object and its member functions and responds appropriately.

HINT!

Quotation marks are used around a header file's name. The angle brackets, < >, are used for files that come with the C++ application that you buy.

EXERCISES

Label each file as a header file or an implementation file.

1. triangle.h
2. geofigure.cpp
3. iostream.cpp

PARTS OF A HEADER FILE

The header file begins with a class definition. The class is defined using a name followed by beginning and closing braces ({   }) followed by a semicolon to conclude the definition of the class. In our example, we define a class called "Address_book." (Think of the class as the "type" for the object.)

The class and its member functions, and information regarding the type of object we are creating can be defined in two ways. The class functions can be separated into two sections: a *public* and a *private* section. Functions defined in the public section are accessible through any client program. If a function is defined in the *private* section, however, it *cannot be accessed* by a client program.

The reason for having both sections is to protect software from being accessible by other programs. The private section of a class allows the programmer to "hide" details from a user. This can be useful when you are trying to keep the client program free of excessive detail. Recall that the reason we have an implementation file is to keep code out of the client program. The other reason is to hide details regarding design from the user as well. The private section lists the functions that only the class, itself, can use.

An outline of a class definition:

```
#include <iostream.h>
class Address_book
{

public: // public member functions are listed here.



private:// private members are listed here.


};
```

HINT!

Functions and members of a class are assumed to be private members unless otherwise specified. However, it is good practice to explicitly list both the public and private sections of a class.

ON THE CD-ROM!

A complete header file and implementation file are shown.

We will revisit the class definition after we discuss the next topic, the *object*.

DESIGNING AN OBJECT

Let's start to define an object of a particular class. We'll start with deciding which data types would be useful for a "person" object. Think of all the information you might like to store in this object. We would need some strings for a person's name, address, and phone number. We will need an integer for the age of the person. There are several other variables we could define but let's stop there.

An *attribute* of an object is similar to a field of a structure or record. It is a particular data type used to represent a part of the object. Objects usually involve more than one of the standard data types we have seen—int, char, double, and so on. As a result, the object has a few different parts, each of which can be a separate type and is called an *attribute*.

After designing the object, we can then decide which functions would be most suitable for the object.

EXAMPLE: ATTRIBUTES OF A "PERSON" OBJECT

If we design an object used to represent the data we want to store for a person, we need three strings, each to hold a name, address, and phone number and then an integer to hold an age.

```
// The object for a person

string  name;
string address;
string phone_num;
int age;
```

Our object has *four* different attributes—three of which are strings and one of which is an integer. Let's consider some other examples of objects used to represent such items as a geometric figure and a list of movies currently playing.

EXAMPLE: OBJECT FOR A GEOMETRIC FIGURE

We want to design an object that will appropriately define some geometric figure. Most figures, such as triangles, squares, rectangles, and the like have a given number of sides and then two additional factors to consider—the lengths of the sides and whether those sides are equal. Let's design an object for a triangle:

```
// The object for a triangle

double side_1;
double side_2;
double side_3;
char sides_equal;
```

EXAMPLE: OBJECT FOR A LIST OF MOVIES CURRENTLY PLAYING

Another example of an object is a list of items. We represent a list as an array that holds strings. Each string is a title of a movie. We will make the list 25 members long but we do not plan to fill all those slots. So, in addition to the list, we will have an integer used to represent the actual number of movies that are in the list.

For example, we might only have 10 movies in the list even though the array can hold 25 movies. So the integer variable (we'll call it movie_ count) holds the value *10*. Here is a definition with two attributes for the object:

```
string movie_list [25];
int movie_count;
```

Picture a long array of which only 10 slots might be used. Each object that we have discussed has more than one attribute (Figure 17.3).

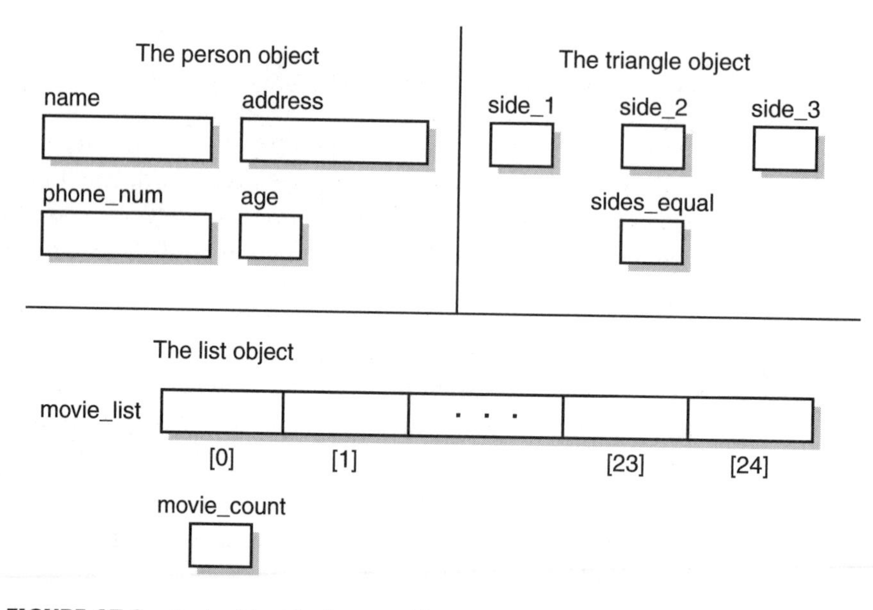

FIGURE 17.3 Each object is shown with its attributes.

17.4 HOW TO DEFINE MEMBER FUNCTIONS OF A CLASS

Let's revisit one of the objects from the previous section. For this object, let's write two functions that would be suitable for a class of that type. In the example for the "person" object, we will design a printing function so that we can see all the attributes' values on the screen. A second function that would be useful would be one that allows us to change the person's address.

Member Functions of the Person Class

```
void Print_Info (    ) ;

void Change_Address (  );
```

Keep in mind these are just the headings for the functions. You might be wondering how we assign data to the object itself. There are special functions that allow an object to be assigned appropriate data. These functions are called *constructors.*

CONSTRUCTORS

A *constructor* is a function that brings the object "into existence," so to speak. The word "instanciate" is used to describe this process. We "declare" variables of a certain type but we "instanciate" objects. This word comes from the phrase, "to give an instance" of the object. Just think of this word as another word to describe the way an object is created.

Constructor functions are functions that assign all the attributes of an object. Within the constructor function, each attribute is used in an assignment statement. Data is assigned by the user or by the programmer, depending on how the code is written.

Let's look at a heading for a constructor function for the "person" object. In the next section, we will show this function containing statements that allow each of the attributes of the object to be assigned. Here is the heading of the constructor:

```
person ( );
```

There are a couple of things to notice about the heading for a constructor. First, there is no return type listed—not even the word *void.* Second, the name of the constructor is *always* the name of the *object type,* that is, the name of *the class itself.*

EXERCISES

Write the headings of the constructors (functions) for each of the following objects.

1. The triangle object
2. The list object

A class may have more than one constructor for an object. There are constructors that assign attributes in different ways. Some constructors use input from the user to assign the attributes whereas other constructors use values selected by the programmer. Another useful constructor is one that allows an object to have its attributes set with the values that are *in another object*. This constructor is called a *copy constructor* because it allows the contents of one object to be copied into another object (Figure 17.4).

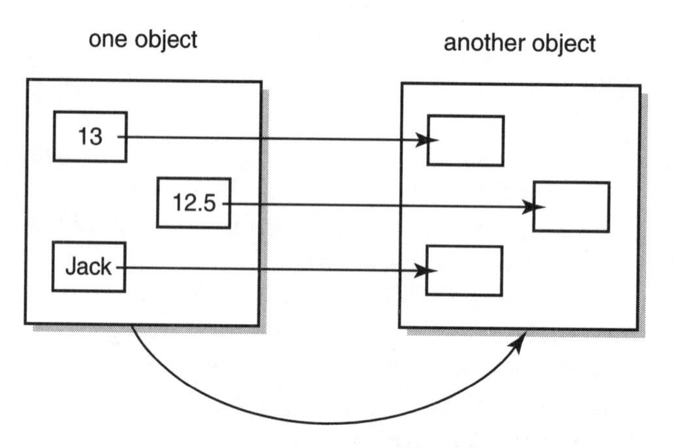

FIGURE 17.4 Two objects are shown—the object on the left copies its values into the object on the right.

17.5 CONSTRUCTORS: HOW OBJECTS ARE INSTANCIATED

The client program uses the class by calling on the class member functions available to it (i.e., the functions that are in the public section of the class). By keeping all the code for the object and its methods in a separate file, the client program is an easier program to read. The client program is an example of the interface between the user and the programmer (the client program and the class itself). It is an example of how one program talks to another.

All of the detailed work that the program accomplishes is carried out by calls from an object, instanciated (declared) in the client program, to member functions of the class.

Let's examine two different kinds of constructors.

CONSTRUCTORS WITH NO PARAMETER LIST

Here are examples of constructor functions for some of the objects in the previous sections. Each constructor is written without a parameter list. Only the attributes are assigned, either by the programmer or the user.

```
person( )
{
name = " Vanessa Vardwick";
address = "114 Plumberly Road"
phone_num = " 234-5678 ";
age = 16;
}

triangle  (       )
{
side_1= 12.0; // the programmer assigns
side_2 = 5;
side_3 = 13.0;
sides_equal = "n";
}
```

CONSTRUCTORS WITH A PARAMETER LIST

Now we will rewrite each of the constructors using a parameter list, where we expect some values to come from another source, such as the client program.

```
person(string nam1, string addl, string phonl, int yrl)
{
name = nam1;
address = add1;
phone_num = phon1;
age = yr1;
}
```

```
triangle  (double s1, double s2, double s3, charm)
{
side_1= s1; // the programmer assigns with values sent from
                        //the client program.
side_2 = s2;
side_3 = s3;
sides_equal = m;
}
```

HINT!

Constructors have *no* return type—not even the word *void.*

CALLING THE CONSTRUCTOR FROM WITHIN THE CLIENT PROGRAM

Once we have written the class (including the constructor functions mentioned) we are ready to call the constructors from the client program. A call to a constructor will look almost exactly like a declaration of a variable. Here is a call to each of the constructors mentioned previously.

```
person      my_friend ( );
```
object type   object name

```
triangle    tri_fig ( );
```
object type   object name

Now let's do some examples where we call a constructor that has a parameter list for each of the objects mentioned. The only difference in these calls will be that we send in parameters to each constructor call. This time you will notice that the parentheses to the right of each object name are filled with variables that have been assigned values.

EXAMPLES

I.

```
string nam, add, phon; int yr;
cout<< "Please type your name, address and phone number.\n";
cin >> nam;
```

```
cin >> add;
cin >> phon;
cout << "Please type your age.\n";
cin >> yr;

person      my_friend (nam,add,phon,yr );
```

object type object name

II.

```
double sd1,sd2,sd3;
sd1 = 5.9;
sd2 = 4.2;
sd3 = 4.7;
char side_equal = "n";
triangle tri_fig (sd1,sd2,sd3,side_equal );
```

object type object name

Here is a short example of a class header and implementation file. Notice that in the implementation file, you will see a colon used twice (::). This is called a scope resolution operator. On one side of the operator is the class name and on the other side is the member function name.

THE SCOPE RESOLUTION OPERATOR

The *scope resolution operator* is used so that a member function will be associated with a given class. In the example that follows, note the heading of the print function in the implementation file.

```
void person:: Print_Info()
```

void	person	::	Print_Info()
return typ	class name	scope operator	member function name

This operator (::) is used to identify a function as a member of a certain class. It is used to clarify this relationship. Many classes probably contain a printing function with the same name. By using the scope resolution operator, we identify this particular function as the "printing" function that belongs to the "person" class and not another printing function of another class.

```
//the header file
class person
```

```
{
public:
person ( );
void Print_Info (   ) ;
void Change_Address (string address2  );
void Change_Age ( int year);

private:

string name, address, phone_num;
int age;
};

// the implementation file
person::person ( )          // a constructor
{
cin >> name;
cin >> address;
cin >> phone_num;
cin >> age;
}
void person:: Print_Info()
{
cout << "Name : " << name << endl;
cout << "Address : " << address << endl;
cout << "Phone Number : " << phone_num << endl;
cout << "Age : " << age << endl;
}
void person::Change_Address( string address2)
{
//replacing one address with the parameter given
address = address2;
}

void person::Change_Age( int year)
{
age = year;
}
```

ON THE CD-ROM!

These two files along with a client program are used to show the relationship among all three files.

SUMMARY

First we explained that C++ is an "upgraded " version of the C language; it uses objects and is an object-oriented language. Objects are very similar to structs, which we examined in Chapter 10. In object-oriented programming, the object is the focus of the programming. As a result, the object makes all calls to functions within the class.

Objects have different parts, each of which is called an attribute. Objects are assigned through calls to a special function called a constructor. When the constructor is called, we say that the object has been "instanciated." This means that an instance of the object has been created. Some constructors use parameter lists while others do not.

A class is defined as the grouping of an object, its attributes, and the functions designed to work on the object. These functions include the constructors we just mentioned and other functions that might alter some attribute of the object, display the attributes of the object on the screen, and so on.

A class is represented through two files—the *header* file and *implementation* file. The header file is where an overview of the class is listed. Only the function headings are listed here as well as the attributes of the object and their type. The implementation file contains the entire code for each function in the class. A third file, called a client file, is the file that accesses the class member functions.

The header file may be separated into two sections: a public section and a private section. The public section contains a list of all functions that can be accessed through the client program. The private section is typically where the object's attributes are listed and any functions that the programmer does not want the client (user) to see.

Header and implementation files are made accessible to the client file through the include statement, which we first saw in Chapter 4.

ANSWERS TO EXERCISES

17.3

1. header file
3. implementation file

17.4

1. triangle ( ) ;
2. list ( ) ;

CHAPTER

18

Java: It Sounds Great But What Does It Do?

IN THIS CHAPTER

- Java's Role in the Internet
- Platform Independence
- The Java Virtual Machine—an Interpreter
- Byte Codes
- Java Classes—the Import Command
- Inheritance and the "Extends" Command
- The Organization of a Java Program
- Defining Components, Containers, and Layout
- Defining Events and Listeners
- Running a Java Program
- Java Applet vs. an Application

18.1 WHY DOES JAVA EXIST?

Languages are created to solve some existing problem or to envision some new way of employing technology. Java solves a real problem with widespread network use. As mentioned in the chapter on HTML, when sending a file through a network to another computer, you may lose some of the original qualities of the file. HTML allows a programmer to deliver a text-based file along with directions that allow the text to be represented correctly, or as intended (Figure 18.1).

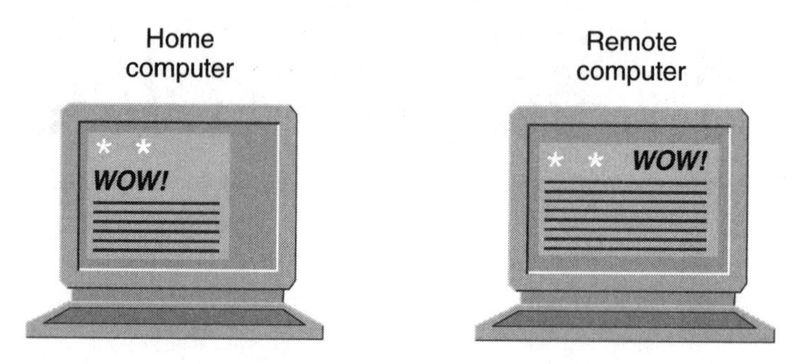

FIGURE 18.1 A file shown on its own computer and what it looks like on a remote computer.

But what about sending something more complicated than text over a network such as the Internet? How can we send a program to a remote computer whose C.P.U. may be different from our own? A program that executes some action like opening a window or moving an object on the screen is more involved than an HTML file. A program has directions in the form of programming statements and is designed to make something happen on the other end—like spinning a loop, executing a decision, and so on. Sending a program to an unknown computer that has different hardware could cause problems in the execution of the program. Java is the answer to that problem.

PLATFORM INDEPENDENCE

Java is a language that is *platform independent*. That is, Java is designed to work on any machine, regardless of its hardware (platform or architecture) or C.P.U. PCs differ from Macs in their machine language because they have different microprocessors. Java was designed to overcome the problem of transmitting a program to machines with different designs.

How this is done will seem a little mysterious at first, but it is, in fact, very clever.

18.2 THE JAVA VIRTUAL MACHINE

Let's review some of the basics of how programs are translated into understandable code. When a program is written in a high level language, it must be translated so that it can be executed by the C.P.U. in instructions that the processor understands. Translators, or more specifically, compilers and interpreters, must be able to understand two things: the high level language in which the program is written (such as C, C++, or B.A.S.I.C.) and the machine language of the processor of the computer.

MORE INFORMATION ABOUT COMPILERS

A *compiler* "reads" an entire program all at once and then translates it into "object" code—machine code understandable by that machine. The object code is stored in a separate file called an "executable " file because it is now ready to be executed. This separate file can be run without being retranslated.

The advantage of a compiled program is that this new file, the "executable" file, can be run on the machine at any time. It does not need to be retranslated *unless the source code is altered* by the programmer. Compiled programs run very quickly.

An *interpreter* translates *line by line*. As the program is run, each line is translated, then executed. Unlike a compiler, which generates the "executable" file, an interpreter does its work *while the program is being run* and saves no copies of that work. For this reason, there is no special file that is saved anywhere. Again, consider the analogy of an interpreter of a spoken language. The interpreter always stands next to the person speaking so he can translate each line just after it is spoken. If you run the program a second time, it will take just as long as it did the first time because the interpreter must do its work all over again, regardless of whether a change was made in the program.

Java was designed to combine the best features of both types of translators. Java translates a program through a compiler to an "intermediate" state. At that point in time, the program has been translated into *byte codes*. Byte codes do not represent the object code that is machine language. These byte codes need to be interpreted for the particular architecture of the receiving machine into "native" language. This last stage of translation is done through a *Java Interpreter*.

THE JAVA INTERPRETER

As long as a computer has a Java interpreter, the Java program can be run on the machine, regardless of where the program originated. This is the big advantage of Java. One of the problems that arose with widespread Internet use was how to send a program over the Internet so that it could execute on the other end—at the remote computer, regardless of its architecture.

By compiling a program into a "semi-translated" state, the language takes advantage of the speed that a compilation process offers. This "semi-translated" state produces *byte codes,* which are then interpreted by a Java Language Interpreter. Presently there are interpreters for all major computer architectures. The architecture refers to the design of a computer system. As long as a computer has a Java Interpreter, it can finish the translation of byte codes into native machine language (Figure 18.2).

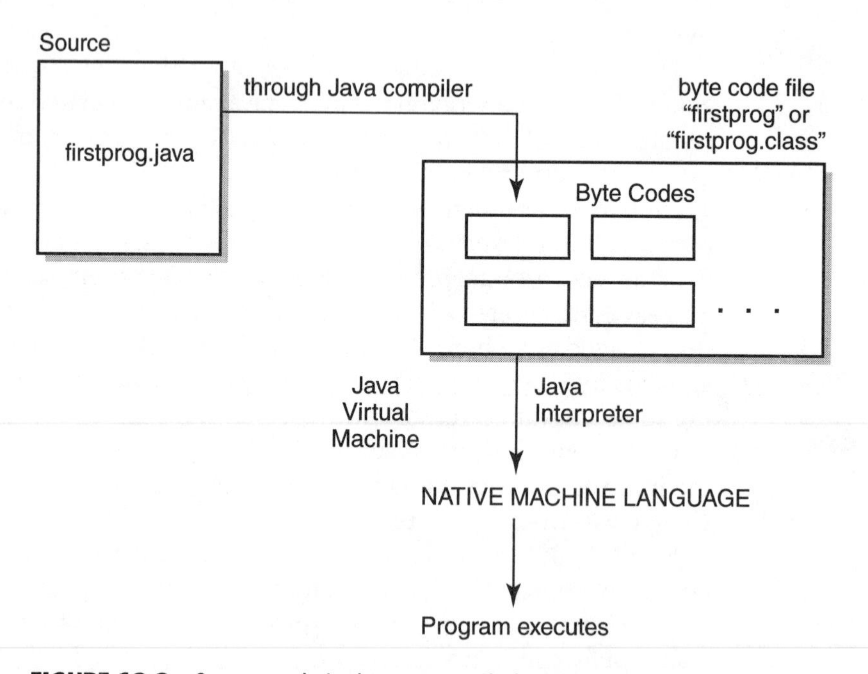

FIGURE 18.2 Source code is shown compiled into byte codes. Byte codes are shown before they are translated into "native" machine language.

Byte codes represent the intermediate state of translation. The Java Compiler generates these byte codes, which are then passed along to the receiving computer. This computer then translates these byte codes into native machine language. This last step in translation is done by a Java Interpreter. As long as your machine has a Java Interpreter, the program can be executed on the "local" machine.

"Write Once, Run Everywhere" is a saying associated with the Java language. It means that you can write one program in Java and any machine will be able to run that program. Although there are some problems associated with the individual interpreters of machines, for the most part this is true.

18.3 LANGUAGE SYNTAX

The Java language uses the same syntax as C++ (which we have been using throughout this book.) It was a good move on the part of the Java language developers to keep it similar to an already existing and useful language. The only difference with Java is that it is a fully object-oriented language.

FULLY OBJECT ORIENTED

Java is a completely object-oriented language. What this means is that all code within a Java program is written with objects as the focus of the syntax. Unlike C++, which can support function calls that do not use objects, Java does not. If you read a typical Java program, you will notice the syntax of calls to constructors and functions defined within a class. These functions that are defined within a class are called *methods* in the Java language. Consider these examples:

An object is instanciated.

```
Jbutton     b            =            new      Jbutton();
```

object   object name   assignment   new command causes a pointer to the object

An object is instanciated.

```
frame fram = new frame();
```

An object calls a method (a function) of the class.

```
fram.setSize   (200,250);
```

object calls setSize with two parameters

When you read a Java program, you will notice methods that call other methods. The first method (really a function) is recognized by the following heading. (This method is analogous to the int main function of C++ because that represents the beginning of program control.)

```
public void static main (String args[])
{
// object is instanciated here
// other methods are called in here
}
```

This method is the beginning point of program organization. From this point, all other methods are activated through the creation of the objects that call them. C++ provided our first glimpse of the organization of a program into classes and the objects that are representations of a class.

SOME JAVA CLASSES

Because Java is fully object oriented, we can expect to use a large group of classes that come with the Java language. These classes provide access to many interesting methods that perform a variety of tasks on objects. Some of these objects are windows, buttons, frames, panels, checkboxes, and the like. These objects are very useful when designing your own Web page.

Two important classes with which you should become familiar are the Java *Abstract Windows Toolkit*, known as A.W.T., and the Java *Swing* class, known as *Javax Swing* (for JAVA 2 SDK version 1.2.1). These two classes will allow you to manipulate windows, the mouse, and the keyboard, to name just a few of the things you can do.

Before you write your first Java program, you need to learn a little more about the manner in which you would create some object before manipulating it. Recall from our chapter on C++, that *constructors* are special functions that every class has, and they are used to "create an instance" of the class—that is, to bring into existence an *object* of that class. When you call a constructor in Java, you will use the new command, which we first saw with the pointer variables from Chapter 12. Here is an example:

```
JRadioButton  first = new JRadioButton();
```

object type   object name   the new command is used

We use the *new* command when constructing an object in Java. This is different from how we called constructors in the C++ programming language (in Chapter 17). The first time we saw the new command was in the chapter on pointers. The *new* command indicates that we are not only instanciating an object, but also assigning a *pointer* to it at the same time.

INCLUDING CLASSES IN YOUR JAVA PROGRAM

To give your program access to the many different classes available in the Java language, we use a special word, called "*import,*" which allows us full access to a class within the program we write. This command is similar to the "*include*" directive we saw in the C++ programming language. Here are two examples where we use the "import" statement to access two different classes. When we import the class, we use the class name and we also use an asterisk ("*") next to the class name to indicate that we want all the methods associated with that class.

```
import java.awt.*;
import javax.swing.*;
```

class name . all the methods will be accessible because of the asterisk

OBJECTS AND INHERITANCE

Because Java uses many existing classes to do some of its tasks, such as responding to clicks of a mouse or displaying a large picture (graphics file) on your screen, you want to always save time in your writing of code by making use of existing classes.

Sometimes you may want to adapt a class and its methods for some purpose that is similar to the methods that the class already provides. Consider this example. You wish to draw a frame on the screen but you want the frame to have four panels inside of it. What you are really doing is *extending* the frame class by making it a frame with some panels you wish to design.

The idea of extension comes from its meaning in the English language. To extend something means to stretch it or push its concept a little further. Consider these commands in the Java language.

```
class largeframe extends frame
class stainglasswindow extends Jwindow
```

The first class, "largeframe" is an extension of an existing class called "frame." The new class "largeframe" will inherit all the methods in the frame class. What this means is that we do not have to write any new methods for the largeframe class if they already exist in the frame class.

The concept of *inheritance* exists in other languages besides Java—namely C++. Think of what the word "inherit" means in English. Here are some examples: "I inherited my mother's ability to read for hours on end." Another example would be "I inherited my father's musical talent." In computer programming languages, *inheritance* implies that a class has all the characteristics of the parent class. The parent class is called the *base* class in Java. Remember that a class consists of an object and all the methods associated with it.

Java uses the word "extends" to indicate the concept of inheritance. When we say that one class "extends" another class, think about the meaning of the word "extension" in English. If one class extends another class, it inherits everything from that first class, called the "base" class. If the base class has three methods associated with it, the new class will have those three methods and any other methods that it defines (Figure 18.3).

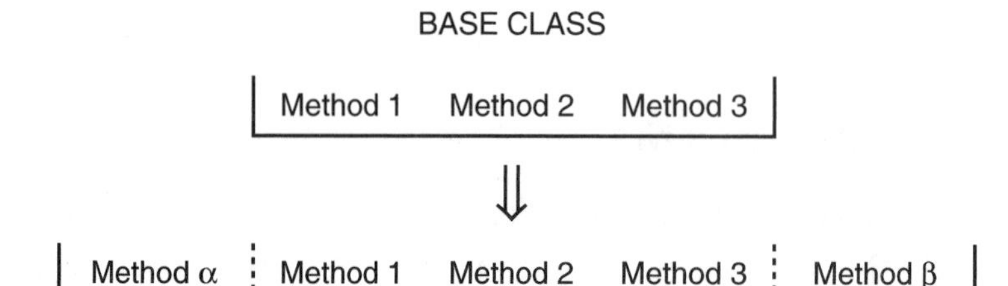

FIGURE 18.3 A new class is shown inheriting all the methods of a base class.

18.4 THE ORGANIZATION OF A JAVA PROGRAM

Java is used extensively in Web page design. A Web page that is used to ask the user to make choices via selecting a button (called a "radio button") or marking a check box could have been written in Java. But what part of the language provides this kind of interaction with the user? So far in our other programs, we have only been able to interact with the user in a limited way by printing comments for the user to read on the screen and waiting for the user to type a response such as a number or string.

Java provides the ability to develop what is known as a *graphical user interface* or G.U.I. for short. This means that the user interacts with the program via prompts that are displayed as graphics rather than just text. Think of your desktop on a Mac or a PC. Most computers use a G.U.I. as a way of communicating with the user. Graphics that are used on a screen to invite some kind of response from the user (the viewer) represent the meaning of the term *graphical user interface*. So instead of simply *asking* the user a question and waiting for his *typed* response, the program waits for the user to do something like click a button or a check box.

A typical Java program designed to create a G.U.I. will show a window with some boxes marked with text. Then the user can decide which boxes he will mark by clicking the button or marking a checkbox. Once he clicks the box, which we call an "event," the computer registers that event and issues an appropriate response. How Java controls these "events" will be the subject of the next section.

COMPONENTS AND CONTAINERS

Java uses two words to describe how a window is designed on the screen. Components are part of a window, or rather they are elements that are put inside a window. Think of all the elements that you have seen in a window at a Web site. You have probably seen checkboxes, radio buttons, panels, and the like. Each of these elements is called a *component* because it can be stored inside of a larger element called a *container*.

Examples of *containers* are frames and panels. You can put checkboxes inside of frames. You can also put a graphic image inside of a frame. A panel is a smaller section within a frame. You can put components inside of a panel. Because a panel can sit inside of a frame, a panel is both a component of a frame and a container for other components (Figure 18.4).

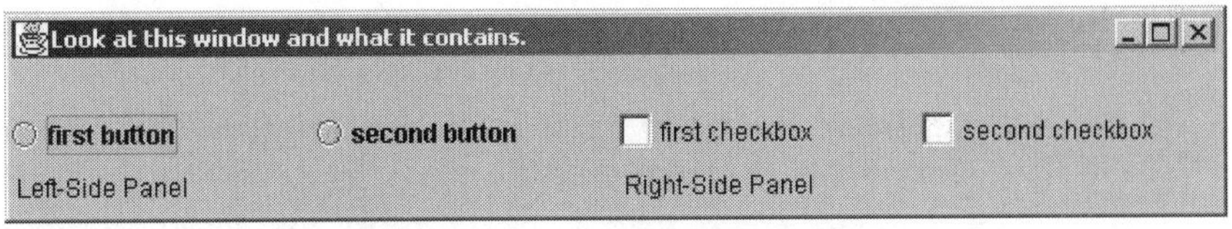

FIGURE 18.4 A frame is shown with various components inside of it. Radio buttons and checkboxes are shown inside of the frame and also a panel.

LAYOUT MANAGEMENT

Components and containers are then organized and arranged according to some layout or design. This is done through calls to another class of methods called the layout class. If you have several buttons within one window, will all the buttons go at the top of the window or will they be arranged in some other manner (Figure 18.5)?

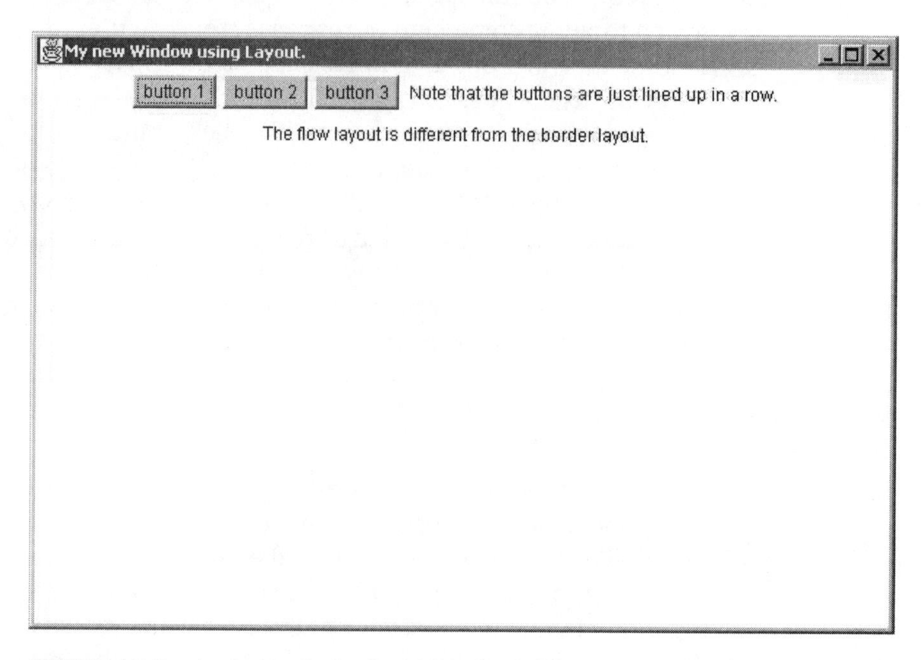

FIGURE 18.5 A window is shown with individual buttons all grouped at the top of the window.

The layout class offers a few different ways of organizing objects. As an example of one kind of layout, consider a grid that has been separated into rows and columns. If the grid layout is declared, then all the objects within the container will be put inside of a grid with the dimensions (rows and columns) you have chosen.

Consider a grid that has two rows and three columns. If you have five components that you wish to place inside of this grid, you could put one in each slot of the grid, starting in the top row and working across each column until you have filled that row. Then you can fill the second row with the other components (Figure 18.6).

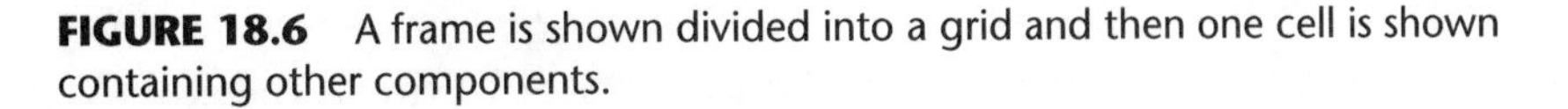

FIGURE 18.6 A frame is shown divided into a grid and then one cell is shown containing other components.

EVENTS AND LISTENERS IN JAVA

For a graphical user interface to work, the computer has to be able to respond to some action that has occurred—that is, an event that happened on the screen, like a mouse being clicked. If the user clicks the mouse on a checkbox in a window, the "clicking" represents an event that occurred on the checkbox and a listening method (function) has to be invoked so that the event can be processed. This is very different from a computer having to recognize whether an answer typed at the keyboard has the same value as some variable's value. Certain functions are called "listening" functions because they "listen to" some event (an action) that occurred on an object .

HOW TO RUN A JAVA PROGRAM

Whether you use Java on a Windows platform or a Mac platform, you need to be aware of the different stages of compilation, especially if you are writing your own programs in Java. The first stage of compilation involves finding the Java compiler and directing it to your source program (source file).

First you need to write a Java program and save it as a Java file. You can use any text editor—you don't need to use a word processor if you have a simple text editor. Once you write the code in Java, you save it as a filename.java file. Now the computer will recognize the extension as a Java file. (Recall that extensions are a few letters that appear after the file name to indicate what kind of file it is.) Let's start with an example:

EXAMPLE

```
Firstprog.java   // this is the name of the source file
```

The next step is to direct the compiler to the file. Identify your java compiler, perhaps via a name like javac. Then direct the compiler to compile "Firstprog.java." After the compiler finishes that task, a file of byte codes will be produced. This new file may just be called "Firstprog." This file does not have an extension (an appendix to the file name). The next step is to identify the Java *Interpreter* (also known as the Java *Virtual Machine*) and to direct the interpreter to finish the translation of the byte code file so that the program can be executed (Figure 18.7).

After the interpreter finishes its work, the program is executed on the machine. The advantage of having this two step process in translation is

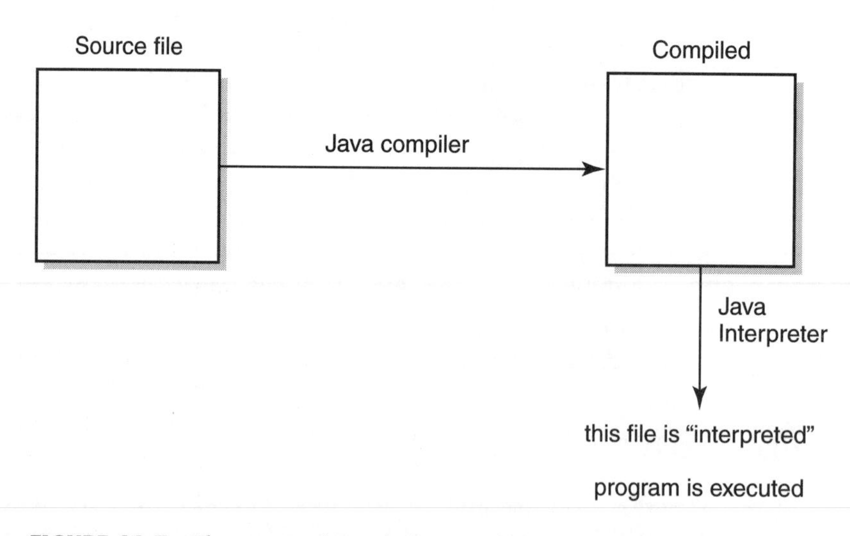

FIGURE 18.7 The stages of translation are shown from first compiling the program into an intermediate file and then the interpreter finishes the translation.

how browsers on the World Wide Web—such as *Netscape* and *Internet Explorer*—are able to bring interesting Web pages into the environment of your home computer. The browser will download a file that was written partially in Java and HTML (as an example) and it will be executed on your machine regardless of where the program was originally written.

Imagine the problem of having files written on machines that are different from our own being unable to execute on a local computer. The Internet would hardly be what it is today without a language like Java, which facilitates communication between different (heterogenous) computers.

Remember that an interpreted file does not exist because it only lasts for the duration of the execution. That is the nature of interpreted code. Compiled code is generated as a separate file. If any changes are made to a file that has been compiled, a new compiled version will be generated—even if only one change is made to the original program. That's the bad part about compilation. The good part is that once the compiled program has been generated, it executes very quickly because it is always ready to be executed.

18.5 JAVA APPLETS

You might have heard the term "applets." It seems like a shortened word for "application." Remember that an application is a set of programs that are able to do some particular task on the computer—such as a word processing application. An *applet* is a small dependent application that executes some particular task. An applet is not a stand-alone application.

It is designed to be executed *from within an application file*. Most java applets that you have probably seen are activated through an HTML file. Remember what the HTML file does? It allows text files to be transmitted so that their formatting is retained at the target computer's level. The formatting is controlled through tags that specify how the text should look.

Java applets could be something like a small animated figure that moves across your screen while you view a Web page. Another java applet might be an image that fades in and out in the corner of a Web page or an applet that takes your input and processes that input. The applets are designed to be executed when called by the larger file. The way to access an applet in an HTML file is through the designated "APPLET" tag.

THE APPLET TAG

When you wish to house a Java applet within an HTML file, you simply use the applet tag provided in the HTML language. Just like any tags we have used in HTML, there is both an opening and a closing tag.

```
<APPLET> // opening tag

</APPLET>// closing tag
```

Within the opening applet tag, you state the name of the java applet file with the appropriate extension, which distinguishes this file as an applet file that has been compiled already. You do not use the ".java" extension, which is reserved for stand-alone Java programs (known as Java applications). You use the ".class" extension.

EXAMPLE

```
<APPLET CODE = "Firstprog.class"  WIDTH = 200 HEIGHT = 100> //
  opening tag

</APPLET>// closing tag
```

Notice that the name of the file uses the extension, ".class."

When the HTML file is executed, it will find the java applet and cause its execution. This is how programmers are able to develop such interesting Web pages. Here is an example of an HTML file that activates a Java applet, called "littleapp":

```
<HTML>
<HEAD>
<TITLE>   Running an applet for the first time <\TITLE>
<\HEAD>
<BODY>
<APPLET  CODE = "littleapp.class"  WIDTH = 200 HEIGHT = 100> //
opening tag
</APPLET>// closing tag
<\BODY>
<\HTML>
```

ON THE CD-ROM

Some short programs and applets in Java.

SUMMARY

Java's role in the Internet is extensive because of its powerful programming capabilities and its ability to be translated on a variety of machines. When a Java program is written, it is first compiled into a semi-translated state called *byte codes*. These codes do not represent machine language but, rather, an intermediate state that needs to undergo further translation to be understood by the native machine. So the final step in translation is completed by a Java *Interpreter* (the Java *Virtual Machine*) that interprets these byte codes and then executes them.

Java is *platform independent* and this describes its ability to run on different machines.

Java uses objects; we saw these first introduced in the C++ programming language. But Java is different in that it is completely object oriented. All code is written with objects as the focus of the syntax. Java *methods* are class member functions. There are many useful classes in Java that can be used in any new program you write. We use the "import" command to access these classes just as we used the "include" directive from C++. Java syntax is just like the syntax of the C++ programming language.

We introduced different types of objects used to develop a G.U.I. (graphical user interface). These objects are called *components* and *containers*. Components, such as radio buttons and checkboxes, can be arranged within containers, such as frames or panels, to create a typical interface on a Web page. The *layout* class allows the programmer to arrange components in some kind of order within a container.

Events and *listeners* are methods that represent a way of responding to user input that is action oriented. If the user *clicks* the mouse on a radio button, for example, an "event" has occurred and a listening method has to be invoked on the button so that it can respond to this event. Then the program can note the user's response to the button.

A Java *applet* is a small Java program that is launched within another application. It is not a "stand-alone" application like a Java program. An applet is typically launched within an HTML file.

APPENDIX

A

Career Opportunities

There are many careers for programmers. Programming allows one to develop a concrete appreciation for how the computer "handles" problems. It is also the first step for careers in anything related to the software development cycle.

There are different kinds of programming languages to learn. Once you have an expertise in one language, you will be able to learn other languages. Programmers who develop their skills against the background of the software development cycle may eventually wish to learn more about databases, operating systems, or how Web sites work.

There are two main categories of programmers. Those who program exclusively and those who determine the design of the software being developed.

Software developers, software engineers, or programmer analysts Programmers who implement (write) the code to fit a project design as specified by a systems engineer.

Systems engineers Programmers who deal with a customer to develop appropriate software based on client need. They are responsible for the design and implementation of software throughout the development cycle.

In addition to these two main categories there are the following opportunities that a programmer might pursue.

Database Architects/Engineers Responsible for the architecture (design) of a relational database for the management of data.

Managing data for the individual needs of an organization is one of the most important uses of the computer.

Web Authors Web authors design and develop Web sites and would use many of the Web-based languages.

GAME PROGRAMMERS

Game programmers are software developers who take the ideas, art, and music and combine them into a software project. Programmers obviously write the code for popular games (e.g., *DOOM, SuperMario, Madden NFL Football*, etc.), but may also have several additional responsibilities. For instance, if an artist is designing graphics for the game, the lead programmer could be responsible for the development of a custom set of tools for the creation of the graphics. It is also their job to keep everything running smoothly and to somehow figure out a way to satisfy everyone from the producer of the game to the artists.

The game programmer is responsible for taking the vast number of elements and combining them to form the executable program. They decide how fast a player can run and how high she can jump. They are responsible for accounting for everything inside of the game's "virtual" world. While doing all of this, they often will attempt to create software that can be reusable for other projects and spend a great deal of time optimizing the code to make it run as fast as possible.

Sometimes, a given project may have several programmers who specialize in one key area such as graphics, sound, or artificial intelligence (AI). The following list details the various types of programmers and what they are primarily responsible for:

Engine or graphics programmers. They create the software that controls how graphics and animations are stored and ultimately displayed on the screen.

AI programmers. They create a series of rules that determine how enemies or characters will react to game situations and attempt to make them act as realistically as possible.

Sound programmers. A sound programmer will work with the audio personnel to create a realistically sounding environment.

Tool programmers. As was previously mentioned, programmers will often write software for artists, designers, and sound designers to use within the development studio.

APPENDIX

B

About the CD-ROM

The CD-ROM has files written in three different languages.

Most of the programs mentioned in the book were written in the C++ programming language and have been saved with the file extension, ".cpp". All of the programs from the first 15 chapters were written in C++.

HTML files included on the CD-ROM can be run by any browser like Microsoft Internet Explorer or Netscape Navigator.

Programs written in Java can be run using the Java SDK 1.2 included on the CD-ROM.

For detailed instructions about how to write and run programs in each of the respective languages, consult the following files on the CD ROM:

CppHowTo.doc
HTMLHowTo.doc
JavaHowTo.doc

RUNNING PROGRAMS AND USING COMPILERS

The CD-ROM includes files for running programs and using compilers.

C++ FOLDER

This file will explain how to run the programs written in C++.

All files mentioned in the book are within the C++ folder on the CD-ROM.

The files in the C++ folder include source files (ending in ".cpp") and executable files (ending in ".exe"). The executable files should run on your PC. If you experience any problems, download a compiler and re-compile the source files for your own computer.

In order to write and run your own C++ programs, you first need to install a C++ compiler, if you do not already have one on your computer.

Where to Download a Compiler

There are several free C++ compilers that are available online. Most of them will operate from the command line rather than from a GUI. *You should familiarize yourself with the necessary commands that are needed to move about the command line.*

You can download a free compiler from either of these sites:

http://www.digitalmars.com or http://www.borland.com

The files on the CD-ROM were downloaded and tested with the Digital Mars C++ (for Win 32 8.26) compiler from the digitalmars site.

How to Run the Compiler

Read any directions that are online regarding the operation of the compiler. The following might work if no instructions were provided.

After you have downloaded and installed a C++ compiler, put one of the files with the ".cpp" extension from the C++ folder into the folder where the compiler is. (After you download a compiler, make sure you find out in which folder the compiler is.)

From the command line, make sure that your path name to this folder is correct:

e.g.
c:\ Cplusplus\fm\bin

Then type the compilername followed by the filename.cpp after that path:

e.g.
c:\ Cplusplus\fm\bin compilername filename.cpp

After the file compiles, type the name of the file (without the extension) at the command line.

c:\ Cplusplus\fm\bin filename

The program should run.

HTML FOLDER

This file will explain how to run the programs written in HTML.

The files with the extension ".html" can be run by a browser.
The files with the extension ".txt" can be viewed by any text editor.
The four examples in HTML have the following names:

Ex_One.txt
Ex_Two.txt
Ex._Three.txt
Ex_Four.txt

Ex_One demonstrates font size, face and colors.
Ex_Two demonstrates the use of lists, both ordered and unordered, as well as the Link tag.
Ex_Three and Ex_Four demonstrate the use of images and their alignment.

Although the current practice of developing HTML uses style sheets which I did not discuss, these examples will help you when you later study them.

How to Create an HTML File.

1. Save another copy of one of the existing text files and edit it. Save it with the ".txt" extension if you wish to keep a separate file with the source code of the HTML document. Save the file with the ".html" extension as well.
2. View the HTML file from within a browser like Netscape Navigator or Microsoft Internet Explorer. Or use a word processing application like Microsoft Word to view the text files (the source code) for the HTML files.

JAVA FOLDER

This file will explain how to run the programs written in Java.

The Java files (along with other necessary files) are inside a folder called "Java Files."

Copy them into the BIN folder within the Java SDK 1.2 folder, after you have installed Java SDK 1.2 from the CD-ROM onto your hard drive. (You should be familiar with using the command line for MS DOS.) The reason you copy them into the BIN folder is because the java compiler is located within that folder.

To Run the Programs on the CD-ROM

From the command line operating within the WINDOWS environment,

```
C:\ JDK1.2\BIN
```

type the name of one of the program examples.

```
C:\ JDK1.2\BIN    java  Rectangle
```

How to Write Your Own Programs in Java

You can write a Java program with any text editor. You do not need to write it inside of a word processing application.

Save a copy of a file under a new name with the extension ".java". This file should be saved as text.

How to Run Your Own Programs in Java

From the MS DOS command line opened within the WINDOWS environment,

```
C:\ JDK1.2\BIN
```

type this command:

```
C:\ JDK1.2\BIN  javac  YourProgram.java
```

After it compiles, the command line will reappear. Now type

```
C:\ JDK1.2\BIN  java  YourProgram
```

and your program will run.

How to Run a Java Applet

First compile the Java applet by typing the following on the command line:

```
C:\ JDK1.2\BIN > javac  Yourapplet.java
```

Then create an HTML file similar to Example.html. In it insert the name of the applet with the ".class" extension inside of the applet tag.

e.g. Yourapplet.class

To launch the applet from the command line,

C:\ JDK1.2\BIN

type

C:\ JDK1.2\BIN > appletviewer FileName.html

APPENDIX

C

ASCII Chart

The use of ASCII characters is important in many VB projects. You can use the following chart if you need to look one of them up:

32	Space	53	5	74	J	95	_
33	!	54	6	75	K	96	`
34	“	55	7	76	L	97	a
35	#	56	8	77	M	98	b
36	$	57	9	78	N	99	c
37	%	58	:	79	O	100	d
38	&	59	;	80	P	101	e
39	‘	60	<	81	Q	102	f
40	(	61	=	82	R	103	g
41	)	62	>	83	S	104	h
42	*	63	?	84	T	105	i
43	+	64	@	85	U	106	j
44	,	65	A	86	V	107	k
45	-	66	B	87	W	108	l
46	.	67	C	88	X	109	m
47	/	68	D	89	Y	110	n
48	0	69	E	90	Z	111	o
49	1	70	F	91	[	112	p
50	2	71	G	92	\	113	q
51	3	72	H	93	]	114	r
52	4	73	I	94	^	115	s

116	t	119	w	122	z	125	}
117	u	120	x	123	{	126	~
118	v	121	y	124	|	127	Del

Glossary

ALU The arithmetic logic unit of a computer. Found in the CPU.

Algorithm A finite set of steps to solve a problem.

Alignment The positioning of text or an image within a document.

Allocate Memory Set aside memory for a variable. The amount of memory set aside depends on the variable type.

Append Add onto the end (of a file).

Applet A small application launched through another file. It does not run independently.

Application Programs that are designed to accomplish some task like word processing, balancing a checkbook, or playing a game.

Arithmetic Operator Any of the operators that represent addition, subtraction, multiplication, or division: +, –, *, /, %.

Array A data structure that holds different values of the same type: e.g. an array of integers. Each slot of the array is a unique variable.

Ascending Order Refers to numbers that are listed in increasing value: e.g. 5, 7, 12, 16, 20.

Assignment Giving a variable a value in a program.

Attribute An additional quality given to a tag's function to expand the features of the tag: e.g. the width attribute for an image tag.

Attribute Value A value assigned to an attribute within a tag: e.g. width = 100.

Attributes The "parts" (variable types) that define an object.

Background The window area that is not the foreground.

Binary Operator An operator that takes two operands.

Binary Search Searching an "ordered" data structure (like an array) by "eliminating" one half of the data structure at a time.

Bit Refers to the binary digits (zero or one) used to form a byte.

Boolean Expression An expression that produces the value true or false.

Boolean Type A variable type that holds the value true or false.

Browser An application program (like Netscape Navigator or Microsoft Internet Explorer) that goes over the Internet to find, display, or retrieve files.

Byte A sequence of eight bits.

Byte Codes Semi-translated code generated from compiling a Java source program.

CPU The central processing unit of a computer. It contains the ALU and the control unit.

Case Statement See Switch/Case Statement.

Character One letter, symbol, or digit. Several characters together form a "string."

Cin Stream The name of the input stream in C++.

Class An object together with all the functions that create, access, and modify it.

Client File A file that uses a class. This file may instantiate objects of the class and call on public member functions of the class.

Comment A descriptive remark in a program that is not executed by the translator. Comments help readers of the program to understand the programmer's intentions.

Compiler Directive A direction given to the compiler. In the C++ programming language, the "include" word is a compiler directive.

Compiler Translator that translates an entire program at once.

Component A term in the Java programming language to describe an object that rests within a container: e.g. a radio button.

Conditional Loop A loop that executes upon some condition being true: e.g. a "while" loop.

Constructor A function that brings an object into existence (instantiates it) and assigns its attributes.

Container A term to describe an object that contains other objects: e.g. a frame.

Control Statement A programming statement designed to manipulate the normal sequential execution of programming statements.

Control Unit Found in the CPU. The control unit regulates program flow (i.e. the order of statement execution).

Control Variable A variable that controls the repetition of the *for* loop.

Copy Parameter A parameter that generates a copy of the value of the original variable that was used in "the call."

Counter Statement A statement that increases (usually by 1) the value in a variable.

Cout Stream The name of the output stream in C++.

Data Information that will be used in a program. Data can be numbers, words, or a combination of these.

Data File A file that contains data.

Data Structure A holder for data. Generally more elaborate than the simple data types like integers, reals, doubles, chars, etc.

Deallocate Memory To release memory that was being used (for a variable).

Declaration of a Variable Introducing a variable in a program by stating its type and name (identifier).

Delete The "delete" command releases memory set aside for a dynamic variable. Used with pointer variables.

Depreciated Tag A tag that is no longer part of the most updated version of HTML. It is gradually phased out of general usage.

Descending Order Refers to numbers that are listed in decreasing value: e.g. 14, 11, 9, 6, 3, 1.

Digitization The process of converting non-numeric information into numbers so that a computer can "understand" it.

Do...While Loop A post-test loop; that is, a loop that executes until some condition becomes false.

Dynamic Variable A variable that can be "created" and "destroyed" during program execution.

End-of-file Marker A marker that signals the end of a file.

End-of-line Marker A marker that separates one line of text from another (on a text file).

Event An action taken on an object: e.g. clicking a checkbox. An event is "recognized" by a special method called a "listener."

External Drive A drive outside of the computer used to expand the storage capacity of the computer.

Extraction Operator Operator that "extracts" or pulls values from a stream.

Field A variable type in a record definition.

Fixed Iterative Loop A loop that spins a fixed (known) number of times: e.g. a for loop.

For Loop A loop that executes a fixed number of times.

Foreground The front part of a window. Images drawn by the programmer are part of the foreground.

Function A separate block of code that accomplishes some task.

Function Heading A line of code that includes a function's name, parameter list, and whether it returns a value.

GUI Graphical User Interface. A design that allows users to interact with the computer through visual components rather than text alone.

Global Variable A variable that is recognized everywhere (globally) in a program.

Graphics The topic of design and manipulation of images on a computer.

Hard Drive Internal memory of a computer.

Hardware The physical components of a computer.

Header File In the C++ programming language, a file that defines a class, contains all the headings of functions in that class, and the definition of the object, itself.

High Level Language A programming language (e.g. C, C++, Java, Pascal, Fortran, etc.) that requires more translation than a low-level language.

HTML Hypertext Markup Language: a high level computer language used to format documents (add images, sound clips, etc.) and link them over the Internet.

Identifier The proper name of a variable. The name refers to how the variable is "identified" in the program.

Ifstream A stream type in the C++ programming language. Data coming in from an external file will "travel" through this stream.

If...Statement A control statement that uses a boolean expression (after the word "if") followed by a conclusion.

Implementation File A file that implements (contains the code for) the functions defined in a class header file.

Index Slot or member of an array. Most arrays start with the index 0 rather than 1.

Inheritance The concept that one class can "inherit" all the characteristics of a base class.

Initialization of a Variable Giving a variable its first (initial) value.

Input What the user types at the keyboard.

Input Stream A small channel that feeds from the keyboard.

Insertion Operator Operator that "inserts" values into a stream.

Instantiate To bring an object into existence by constructing it.

Integers Positive or negative numbers that have no fractional part.

Internet The immense interconnected network of other networks.

Interpreter Translator that translates an entire program line by line.

Linear Search Searching a data structure (like an array) one member at a time.

Link Tag A tag that allows one to "jump" to another Web page by clicking the appropriate label.

Listener A method that responds to an event.

Local Variable A variable that is defined within a function.

Logic Operator Any of these words, "and," "or," or "not," or symbols used to represent them: &&, | |, !.

Loop A block of statements that executes a certain number of times or until some condition changes.

Low Level Language A programming language (e.g. assembly language) that requires little translation to execute at machine language level.

Machine Language The native language of an individual computer. Programs must be translated into machine language.

Main Section The main part of the program; where the compiler first goes to execute program code.

Merge Sort A sort that "cuts" an array into halves repeatedly until there are only single elements. Then those elements are "merged" into order.

Method The name in the Java programming language for a class function.

Minimum The smallest value in a list (an array).

Mod Operator An arithmetic operator which produces the remainder from a division problem.

New The "new" command causes memory to be set aside for a dynamic variable. Used with pointer variables.

Object A data structure designed with certain functions in mind that are useful for that data structure.

Object Code The translated program that is ready to "run" on the machine.

Object Oriented Programming Programming a problem while keeping the object as the center of focus and design, rather than keeping an action or task in mind.

Ofstream A stream type in the C++ programming language. Data going out to an external file will "travel" through this stream.

Operand A number on which an action or operation is performed.

Operating System The system programs that allow the computer to operate and perform its elementary functions like creating, saving, and deleting files.

Operators Symbols that cause actions to be performed on numbers and Boolean or logical values. There are different kinds of operators - arithmetic, relational, and logic.

Output Any data or messages generated and displayed on the screen.

Output Stream A small channel that goes out to the screen.

Parameter List The list of variables and their types required by a function.

Platform Independent Hardware independent. Generally refers to programs (usually Java) that can execute on any machine regardless of architecture.

Pointer Variable A variable that contains a memory address of another variable.

Post-test Loop A loop whose condition is checked after executing the body of the loop.

Pre-test Loop A loop whose condition is checked prior to entering the body of the loop.

Precedence The ranking of an operator.

Private A section of a class definition. Anything defined within this section cannot be accessed by a client file.

Program Flow The sequential execution of programming commands in a language.

Program Fragment A portion or section of a program.

Programmer The person who writes the program.

Programming Solving a problem by designing an algorithm and expressing that algorithm in a programming language.

Protocol How a computer "communicates" with another computer: e.g. hypertext transfer protocol (http), file transfer protocol (ftp), etc.

Public A section of a class definition. Anything defined within this section is accessible in a client file.

RAM Random Access Memory. Also known as volatile memory. It will be lost once the computer is turned off.

ROM Read Only Memory. Memory that is "hard wired" into the computer when the computer is made (i.e. permanent memory).

Reals Numbers that are represented by decimals. In computer languages, these are numbers that are not integers.

Record Also known as a struct in the programming language C++. A data structure that is designed to hold different types of data: e.g. an integer and a string.

Recursion A method of solving a problem by using a "smaller case" of the same problem.

Reference Parameter A parameter that refers back to the original variable used in the call. This variable will be altered if the parameter is altered through the function's code.

Relational Operator Any of the operators that allow comparison between two operands: e.g. <, >, <=, >=, ==, !=.

Reserved Words Words that are reserved because they have a special use in a programming language. They may not be used as variable identifiers.

Resource The file you wish to access from the host computer.

Return Statement A statement in the C++ programming language. It allows a value to be returned to the place where the call to the function was made.

Return Type The type of the return value sent back to the place that called the function. The return type is listed in the function heading.

Return Value A value returned after a function has been executed.

Scope The extent of recognition of a variable. A variable's scope is limited to the function in which it is defined unless it is a global variable.

Searching The process of looking for a value in a data structure (usually searching through an array).

Selection Sort A sort that repeatedly selects the minimum value from a subsection of a list and then puts that value into a specific position in the list (the array).

Software Programs that run on computers.

Sorting The arrangement of data into either ascending or descending order.

Source Code The original program before it has been translated.

Statements The buildings blocks of a program. The "sentences" in a programming language.

Static Variable A variable that exists for the duration of a program.

String A sequence of characters usually used to represent words.

Switch/Case Statement A statement that allows one of several options to be chosen depending on the value of the variable controlling the statement.

Syntax The grammar of a language (i.e. the rules necessary for the language to make sense).

System Software/Programs Includes the operating system, etc. The programs that allow the computer to perform elementary operations and use applications.

Tag In HTML, a tag is used to mark text in some manner: e.g. a paragraph tag is used to begin a new paragraph of text.

Text File A file that stores only text or characters.

Translators Programs that translate high level languages into machine language.

Type The classification of data. Some common types are integers, real numbers, Booleans, characters, and strings.

Unary Operator An operator that takes one operand.

URL Universal Resource Locator; an address of a resource (a file). A URL has three parts: protocol, host, and file name.

User The person who interacts with a program.

Value Parameter A parameter that generates a copy of the value of the original variable that was used in "the call."

Variable Parameter A parameter that refers back to the original variable used in the call. This variable will be altered if the parameter is altered through the function's code.

Variable A Holder for data. Variables are given names to identify them in a program.

Vector Graphics The creation of images through lines that have length and direction.

Void Term used to indicate that a function will not return a value after it has been executed.

While...Loop A pre-test loop; that is, a loop that executes while some condition is true.

Index

N

O

P

S

T

Java disclaimer

Sun Microsystems, Inc. Binary Code License Agreement

READ THE TERMS OF THIS AGREEMENT AND ANY PROVIDED SUPPLEMENTAL LICENSE TERMS (COLLECTIVELY "AGREEMENT") CAREFULLY BEFORE OPENING THE SOFTWARE MEDIA PACKAGE. BY OPENING THE SOFTWARE MEDIA PACKAGE, YOU AGREE TO THE TERMS OF THIS AGREEMENT. IF YOU ARE ACCESSING THE SOFTWARE ELECTRONICALLY, INDICATE YOUR ACCEPTANCE OF THESE TERMS BY SELECTING THE "ACCEPT" BUTTON AT THE END OF THIS AGREEMENT. IF YOU DO NOT AGREE TO ALL THESE TERMS, PROMPTLY RETURN THE UNUSED SOFTWARE TO YOUR PLACE OF PURCHASE FOR A REFUND OR, IF THE SOFTWARE IS ACCESSED ELECTRONICALLY, SELECT THE "DECLINE" BUTTON AT THE END OF THIS AGREEMENT.

1. LICENSE TO USE. Sun grants you a non-exclusive and non-transferable license for the internal use only of the accompanying software and documentation and any error corrections provided by Sun (collectively "Software"), by the number of users and the class of computer hardware for which the corresponding fee has been paid.

2. RESTRICTIONS Software is confidential and copyrighted. Title to Software and all associated intellectual property rights is retained by Sun and/or its licensors. Except as specifically authorized in any Supplemental License Terms, you may not make copies of Software, other than a single copy of Software for archival purposes. Unless enforcement is prohibited by applicable law, you may not modify, decompile, or reverse engineer Software. You acknowledge that Software is not designed, licensed or intended for use in the design, construction, operation or maintenance of any nuclear facility. Sun disclaims any express or implied warranty of fitness for such uses. No right, title or interest in or to any trademark, service mark, logo or trade name of Sun or its licensors is granted under this Agreement.

3. LIMITED WARRANTY. Sun warrants to you that for a period of ninety (90) days from the date of purchase, as evidenced by a copy of the receipt, the media on which Software is furnished (if any and if provided by Sun) will be free of defects in materials and workmanship under normal use. Except for the foregoing, Software is provided "AS IS". Your exclusive remedy and Sun's entire liability under this limited warranty will be at Sun's option to replace Software media or refund the fee paid for Software, if any.

4. DISCLAIMER OF WARRANTY. **UNLESS SPECIFIED IN THIS AGREEMENT, ALL EXPRESS OR IMPLIED CONDITIONS, REPRESENTATIONS AND WARRANTIES, INCLUDING ANY IMPLIED WARRANTY OF MERCHANTABILITY, FITNESS FOR A PARTICULAR PURPOSE OR NON-INFRINGEMENT ARE DISCLAIMED, EXCEPT TO THE EXTENT THAT THESE DISCLAIMERS ARE HELD TO BE LEGALLY INVALID.**

5. LIMITATION OF LIABILITY. **TO THE EXTENT NOT PROHIBITED BY LAW, IN NO EVENT WILL SUN OR ITS LICENSORS BE LIABLE FOR ANY LOST REVENUE, PROFIT OR DATA, OR FOR SPECIAL, INDIRECT, CONSEQUENTIAL, INCIDENTAL OR PUNITIVE DAMAGES, HOWEVER CAUSED REGARDLESS OF THE THEORY OF LIABILITY, ARISING OUT OF OR RELATED TO THE USE OF OR INABILITY TO USE SOFTWARE, EVEN IF SUN HAS BEEN ADVISED OF THE POSSIBILITY OF SUCH DAMAGES.** In no event will Sun's liability to you, whether in contract, tort (including negligence), or otherwise, exceed the amount paid by you for Software under this Agreement. The foregoing limitations will apply even if the above stated warranty fails of its essential purpose.

6. Termination. This Agreement is effective until terminated. You may terminate this Agreement at any time by destroying all copies of Software. This Agreement will terminate immediately without notice from Sun if you fail to comply with any provision of this Agreement. Upon Termination, you must destroy all copies of Software.

7. Export Regulations. All Software and technical data delivered under this Agreement are subject to US export control laws and may be subject to export or import regulations in other countries. You agree to comply strictly with all such laws and regulations and acknowledge that you have the responsibility to obtain such licenses to export,re-export, or import as may be required after delivery to you.

8. U.S. Government Restricted Rights. If Software is being acquired by or on behalf of the U.S. Government or by a U.S. Government prime contractor or subcontractor (at any tier), then the Government's rights in Software and accompanying documentation will be only as set forth in this Agreement; this is in accordance with 48 CFR 227.7201 through 227.7202- 4 (for Department of Defense (DOD) acquisitions) and with 48 CFR 2.101 and 12.212 (for non-DOD acquisitions).

9. Governing Law. Any action related to this Agreement will be governed by California law and controlling U.S. federal law. No choice of law rules of any jurisdiction will apply.

10. Severability. If any provision of this Agreement is held to be unenforceable, this Agreement will remain in effect with the provision omitted, unless omission would frustrate the intent of the parties, in which case this Agreement will immediately terminate.

11. Integration. This Agreement is the entire agreement between you and Sun relating to its subject matter. It supersedes all prior or contemporaneous oral or written communications, proposals, representations and warranties and prevails over any conflicting or additional terms of any quote, order, acknowledgment, or other communication between the parties relating to its subject matter during the term of this Agreement. No modification of this Agreement will be binding, unless in writing and signed by an authorized representative of each party.

For inquiries please contact: Sun Microsystems, Inc. 901 San Antonio Road, Palo Alto, California 94303

JAVATM 2 SOFTWARE DEVELOPMENT KIT STANDARD EDITION VERSION 1.3 SUPPLEMENTAL LICENSE TERMS

These supplemental license terms ("Supplemental Terms") add to or modify the terms of the Binary Code License Agreement (collectively, the "Agreement"). Capitalized terms not defined in these Supplemental Terms shall have the same meanings ascribed to them in the Agreement. These Supplemental Terms shall supersede any inconsistent or conflicting terms in the Agreement, or in any license contained within the Software.

1. Internal Use and Development License Grant. Subject to the terms and conditions of this Agreement, including, but not limited to, Section 2 (Redistributables) and Section 4 (Java Technology Restrictions) of these Supplemental Terms, Sun grants you a non-exclusive, non-transferable, limited license to reproduce the Software for internal use only for the sole purpose of development of your JavaTM applet and application ("Program"), provided that you do not redistribute the Software in whole or in part, either separately or included with any Program.

2. Redistributables. In addition to the license granted in Paragraph 1above, Sun grants you a non-exclusive, non-transferable, limited license to reproduce and distribute, only as part of your separate copy of JAVA(TM) 2 RUNTIME ENVIRONMENT STANDARD EDITION VERSION 1.3 software, those files specifically identified as redistributable in the JAVA(TM) 2 RUNTIME ENVIRONMENT STANDARD EDITION VERSION 1.3 "README" file (the "Redistributables") provided that: (a) you distribute the Redistributables complete and unmodified (unless otherwise specified in the applicable README file), and only bundled as part of the JavaTM applets and applications that you develop (the "Programs:); (b) you do not distribute additional software intended to supersede any component(s) of the Redistributables; (c) you do not remove or alter any proprietary legends or notices contained in or on the Redistributables; (d) you only distribute the Redistributables pursuant to a license agreement that protects Sun's interests consistent with the terms contained in the Agreement, and (e) you agree to defend and indemnify Sun and its licensors from and against any damages, costs, liabilities, settlement amounts and/or expenses (including attorneys' fees) incurred in connection with any claim, lawsuit or action by any third party that arises or results from the use or distribution of any and all Programs and/or Software.

3. Separate Distribution License Required. You understand and agree that you must first obtain a separate license from Sun prior to reproducing or modifying any portion of the Software other than as provided with respect to Redistributables in Paragraph 2 above.

4. Java Technology Restrictions. You may not modify the Java Platform Interface ("JPI", identified as classes contained within the "java" package or any subpackages of the "java" package), by creating additional classes within the JPI or otherwise causing the addition to or modification of the classes in the JPI. In the event that you create an additional class and associated API(s) which (i) extends the functionality of a Java environment, and (ii) is exposed to third party software developers for the purpose of developing additional software which invokes such additional API, you must promptly publish broadly an accurate specification for such API for free use by all developers. You may not create, or authorize your licensees to create additional classes, interfaces, or subpackages that are in any way identified as "java", "javax", "sun" or similar convention as specified by Sun in any class file naming convention. Refer to the appropriate version of the Java Runtime Environment binary code license (currently located at http://www.java.sun.com/jdk/index.html) for the availability of runtime code which may be distributed with Java applets and applications.

5. Trademarks and Logos. You acknowledge and agree as between you and Sun that Sun owns the Java trademark and all Java-related trademarks, service marks, logos and other brand designations including the Coffee Cup logo and Duke logo ("Java Marks"), and you agree to comply with the Sun Trademark and Logo Usage Requirements currently located at http://www.sun.com/policies/trademarks. Any use you make of the Java Marks inures to Sun's benefit.

6. Source Code. Software may contain source code that is provided solely for reference purposes pursuant to the terms of this Agreement.

7. Termination. Sun may terminate this Agreement immediately should any Software become, or in Sun's opinion be likely to become, the subject of a claim of infringement of a patent, trade secret, copyright or other intellectual property right.

License Agreement: Forte for Java, release 3.0, Community Edition, English

To obtain Forte for Java, release 3.0, Community Edition, English, you must agree to the software license below.

Sun Microsystems, Inc. Binary Code License Agreement

READ THE TERMS OF THIS AGREEMENT AND ANY PROVIDED SUPPLEMENTAL LICENSE TERMS (COLLECTIVELY "AGREEMENT") CAREFULLY BEFORE OPENING THE SOFTWARE MEDIA PACKAGE. BY OPENING THE SOFTWARE MEDIA PACKAGE, YOU AGREE TO THE TERMS OF THIS AGREEMENT. IF YOU ARE ACCESSING THE SOFTWARE ELECTRONICALLY, INDICATE YOUR ACCEPTANCE OF THESE TERMS BY SELECTING THE "ACCEPT" BUTTON AT THE END OF THIS AGREEMENT. IF YOU DO NOT AGREE TO ALL THESE TERMS, PROMPTLY RETURN THE UNUSED SOFTWARE TO YOUR PLACE OF PURCHASE FOR A REFUND OR, IF THE SOFTWARE IS ACCESSED ELECTRONICALLY, SELECT THE "DECLINE" BUTTON AT THE END OF THIS AGREEMENT.

1. LICENSE TO USE. Sun grants you a non-exclusive and non-transferable license for the internal use only of the accompanying software and documentation and any error corrections provided by Sun (collectively "Software"), by the number of users and the class of computer hardware for which the corresponding fee has been paid.

2. RESTRICTIONS. Software is confidential and copyrighted. Title to Software and all associated intellectual property rights is retained by Sun and/or its licensors. Except as specifically authorized in any Supplemental License Terms, you may not make copies of Software, other than a single copy of Software for archival purposes. Unless enforcement is prohibited by applicable law, you may not modify, decompile, or reverse engineer Software. You acknowledge that Software is not designed, licensed or intended for use in the design, construction, operation or maintenance of any nuclear facility. Sun disclaims any express or implied warranty of fitness for such uses. No right, title or interest in or to any trademark, service mark, logo or trade name of Sun or its licensors is granted under this Agreement.

3. LIMITED WARRANTY. Sun warrants to you that for a period of ninety (90) days from the date of purchase, as evidenced by a copy of the receipt, the media on which Software is furnished (if any) will be free of defects in materials and workmanship under normal use. Except for the forego-

ing, Software is provided "AS IS". Your exclusive remedy and Sun's entire liability under this limited warranty will be at Sun's option to replace Software media or refund the fee paid for Software.

4. DISCLAIMER OF WARRANTY. UNLESS SPECIFIED IN THIS AGREEMENT, ALL EXPRESS OR IMPLIED CONDITIONS, REPRESENTATIONS AND WARRANTIES, INCLUDING ANY IMPLIED WARRANTY OF MERCHANTABILITY, FITNESS FOR A PARTICULAR PURPOSE OR NON-INFRINGEMENT ARE DISCLAIMED, EXCEPT TO THE EXTENT THAT THESE DISCLAIMERS ARE HELD TO BE LEGALLY INVALID.

5. LIMITATION OF LIABILITY. TO THE EXTENT NOT PROHIBITED BY LAW, IN NO EVENT WILL SUN OR ITS LICENSORS BE LIABLE FOR ANY LOST REVENUE, PROFIT OR DATA, OR FOR SPECIAL, INDIRECT, CONSEQUENTIAL, INCIDENTAL OR PUNITIVE DAMAGES, HOWEVER CAUSED REGARDLESS OF THE THEORY OF LIABILITY, ARISING OUT OF OR RELATED TO THE USE OF OR INABILITY TO USE SOFTWARE, EVEN IF SUN HAS BEEN ADVISED OF THE POSSIBILITY OF SUCH DAMAGES. In no event will Sun's liability to you, whether in contract, tort (including negligence), or otherwise, exceed the amount paid by you for Software under this Agreement. The foregoing limitations will apply even if the above stated warranty fails of its essential purpose.

6. Termination. This Agreement is effective until terminated. You may terminate this Agreement at any time by destroying all copies of Software. This Agreement will terminate immediately without notice from Sun if you fail to comply with any provision of this Agreement. Upon termination, you must destroy all copies of Software.

7. Export Regulations. All Software and technical data delivered under this Agreement are subject to US export control laws and may be subject to export or import regulations in other countries. You agree to comply strictly with all such laws and regulations and acknowledge that you have the responsibility to obtain such licenses to export, re-export, or import as may be required after delivery to you.

8. U.S. Government Restricted Rights. If Software is being acquired by or on behalf of the U.S. Government or by a U.S. Government prime contractor or subcontractor (at any tier), then the Government's rights in Software and accompanying documentation will be only as set forth in this Agreement; this is in accordance with 48 CFR 227.7201 through 227.7202-4 (for Department of Defense (DOD) acquisitions) and with 48 CFR 2.101 and 12.212 (for non-DOD acquisitions).

9. Governing Law. Any action related to this Agreement will be governed by California law and controlling U.S. federal law. No choice of law rules of any jurisdiction will apply.

10. Severability. If any provision of this Agreement is held to be unenforceable, this Agreement will remain in effect with the provision omitted, unless omission would frustrate the intent of the parties, in which case this Agreement will immediately terminate.

11. Integration. This Agreement is the entire agreement between you and Sun relating to its subject matter. It supersedes all prior or contemporaneous oral or written communications, proposals, representations and warranties and prevails over any conflicting or additional terms of any quote, order, acknowledgment, or other communication between the parties relating to its subject matter during the term of this Agreement. No modification of this Agreement will be binding, unless in writing and signed by an authorized representative of each party.

FORTE(TM) FOR JAVA(TM), RELEASE 3.0, COMMUNITY EDITION SUPPLEMENTAL LICENSE TERMS

These supplemental license terms ("Supplemental Terms") add to or modify the terms of the Binary Code License Agreement (collectively, the "Agreement"). Capitalized terms not defined in these Supplemental Terms shall have the same meanings ascribed to them in the Agreement. These Supplemental Terms shall supersede any inconsistent or conflicting terms in the Agreement, or in any license contained within the Software.

1. Software Internal Use and Development License Grant. Subject to the terms and conditions of this Agreement, including, but not limited to Section 4 (Java(TM) Technology Restrictions) of these Supplemental Terms, Sun grants you a non-exclusive, non-transferable, limited license to reproduce internally and use internally the binary form of the Software complete and unmodified for the sole purpose of designing, developing and testing your Java applets and applications intended to run on the Java platform ("Programs").

2. License to Distribute Software. Subject to the terms and conditions of this Agreement, including, but not limited to Section 4 (Java (TM) Technology Restrictions) of these Supplemental Terms, Sun grants you a non-exclusive, non-transferable, limited license to reproduce and distribute the Software in binary code form only, provided that (i) you distribute the Software complete and unmodified and only bundled as part of, and for the sole purpose of running, your Programs, (ii) the Programs add significant and primary functionality to the Software, (iii) you do not distribute additional software intended to replace any component(s) of the Software, (iv) for a particular version of the Java platform, any executable output generated by a compiler that is contained in the Software must (a) only be compiled from source code that conforms to the corresponding version of the OEM Java Language Specification; (b) be in the class file format defined by the corresponding version of the OEM Java Virtual Machine Specification; and (c) execute properly on a reference runtime, as specified by Sun, associated with such version of the Java platform, (v) you do not remove or alter any proprietary legends or notices contained in the Software, (v) you only distribute the Software subject to a license agreement that protects Sun's interests consistent with the terms contained in this Agreement, and (vi) you agree to defend and indemnify Sun and its licensors from and against any damages, costs, liabilities, settlement amounts and/or expenses (including attorneys' fees) incurred in connection with any claim, lawsuit or action by any third party that arises or results from the use or distribution of any and all Programs and/or Software.

3. License to Distribute Redistributables. Subject to the terms and conditions of this Agreement, including but not limited to Section 4 (Java Technology Restrictions) of these Supplemental Terms, Sun grants you a non-exclusive, non- transferable, limited license to reproduce and distribute the binary form of those files specifically identified as redistributable in the Software "RELEASE NOTES" file ("Redistributables")

provided that: (i) you distribute the Redistributables complete and unmodified (unless otherwise specified in the applicable RELEASE NOTES file), and only bundled as part of Programs, (ii) you do not distribute additional software intended to supersede any component(s) of the Redistributables, (iii) you do not remove or alter any proprietary legends or notices contained in or on the Redistributables, (iv) for a particular version of the Java platform, any executable output generated by a compiler that is contained in the Software must (a) only be compiled from source code that conforms to the corresponding version of the OEM Java Language Specification; (b) be in the class file format defined by the corresponding version of the OEM Java Virtual Machine Specification; and (c) execute properly on a reference runtime, as specified by Sun, associated with such version of the Java platform, (v) you only distribute the Redistributables pursuant to a license agreement that protects Sun's interests consistent with the terms contained in the Agreement, and (v) you agree to defend and indemnify Sun and its licensors from and against any damages, costs, liabilities, settlement amounts and/or expenses (including attorneys' fees) incurred in connection with any claim, lawsuit or action by any third party that arises or results from the use or distribution of any and all Programs and/or Software.

4. Java Technology Restrictions. You may not modify the Java Platform Interface ("JPI", identified as classes contained within the "java" package or any subpackages of the "java" package), by creating additional classes within the JPI or otherwise causing the addition to or modification of the classes in the JPI. In the event that you create an additional class and associated API(s) which (i) extends the functionality of the Java platform, and (ii) is exposed to third party software developers for the purpose of developing additional software which invokes such additional API, you must promptly publish broadly an accurate specification for such API for free use by all developers. You may not create, or authorize your licensees to create, additional classes, interfaces, or subpackages that are in any way identified as "java", "javax", "sun" or similar convention as specified by Sun in any naming convention designation.

5. Java Runtime Availability. Refer to the appropriate version of the Java Runtime Environment binary code license (currently located at http://www.java.sun.com/jdk/index.html) for the availability of runtime code which may be distributed with Java applets and applications.

6. Trademarks and Logos. You acknowledge and agree as between you and Sun that Sun owns the SUN, SOLARIS, JAVA, JINI, FORTE, and iPLANET trademarks and all SUN, SOLARIS, JAVA, JINI, FORTE, and iPLANET-related trademarks, service marks, logos and other brand designations ("Sun Marks"), and you agree to comply with the Sun Trademark and Logo Usage Requirements currently located at http://www.sun.com/policies/trademarks. Any use you make of the Sun Marks inures to Sun's benefit.

7. Source Code. Software may contain source code that is provided solely for reference purposes pursuant to the terms of this Agreement. Source code may not be redistributed unless expressly provided for in this Agreement.

8. Termination for Infringement. Either party may terminate this Agreement immediately should any Software become, or in either party's opinion be likely to become, the subject of a claim of infringement of any intellectual property right.

For inquiries please contact: Sun Microsystems, Inc. 901 San Antonio Road, Palo Alto, California 94303 (LFI#91205/Form ID#011801)

License Agreement: Forte for Java, release 3.0, Enterprise Edition Try and Buy, Multi-Language

To obtain Forte for Java, release 3.0, Enterprise Edition Try and Buy, Multi-Language, you must agree to the software license below.

Sun Microsystems Inc. Try and Buy Binary Software License Agreement

SUN IS WILLING TO LICENSE THE ACCOMPANYING BINARY SOFTWARE IN MACHINE- READABLE FORM, TOGETHER WITH ACCOMPANYING DOCUMENTATION (COLLECTIVELY "SOFTWARE") TO YOU ONLY UPON THE CONDITION THAT YOU ACCEPT ALL OF THE TERMS AND CONDITION CONTAINED IN THIS TRY AND BUY LICENSE AGREEMENT. READ THE TERMS AND CONDITIONS OF THIS AGREEMENT CAREFULLY BEFORE OPENING THE SOFTWARE MEDIA PACKAGE. BY OPENING THE SOFTWARE MEDIA PACKAGE, YOU AGREE TO THE TERMS OF THIS AGREEMENT. IF YOU ARE ACCESSING THE SOFTWARE ELECTRONICALLY, INDICATE YOUR ACCEPTANCE OF THESE TERMS BY SELECTING THE "ACCEPT" BUTTON AT THE END OF THIS AGREEMENT. IF YOU DO NOT AGREE TO ALL THESE TERMS, PROMPTLY RETURN THE UNUSED SOFTWARE TO YOUR PLACE OF PURCHASE FOR A REFUND OR, IF THE SOFTWARE IS ACCESSED ELECTRONICALLY, SELECT THE "DECLINE" BUTTON AT THE END OF THIS AGREEMENT.

LICENSE TO EVALUATE (TRY) THE SOFTWARE: If you have not paid the applicable license fees for the Software, the Binary Code License Agreement ("BCL") and the Evaluation Terms ("Evaluation Terms") below shall apply. The BCL and the Evaluation Terms shall collectively be referred to as the Evaluation Agreement ("Evaluation Agreement").

LICENSE TO USE (BUY) THE SOFTWARE: If you have paid the applicable license fees for the Software, the BCL and the Supplemental Terms ("Supplemental Terms") provided following the BCL shall apply. The BCL and the Supplemental Terms shall collectively be referred to as the Agreement ("Agreement").

EVALUATION TERMS

If you have not paid the applicable license fees for the Software, the terms of the Evaluation Agreement shall apply. These Evaluation Terms add to or modify the terms of the BCL. Capitalized terms not defined in these Evaluation Terms shall have the same meanings ascribed to them in the BCL. These Evaluation Terms shall supersede any inconsistent or conflicting terms in the BCL below, or in any license contained within the Software.

1. LICENSE TO EVALUATE. Sun grants to you, a non-exclusive, non-transferable, royalty-free and limited license to use the Software internally for the purposes of evaluation

only for sixty (60) days after the date you install the Software on your system ("Evaluation Period"). No license is granted to you for any other purpose. You may not sell, rent, loan or otherwise encumber or transfer the Software in whole or in part, to any third party. Licensee shall have no right to use the Software for productive or commercial use.

2. TIMEBOMB. Software may contain a timebomb mechanism. You agree to hold Sun harmless from any claims based on your use of Software for any purposes other than those of internal evaluation.

3. TERMINATION AND/OR EXPIRATION. Upon expiration of the Evaluation Period, unless terminated earlier by Sun, you agree to immediately cease use of and destroy Software.

4. NO SUPPORT. Sun is under no obligation to support Software or to provide upgrades or error corrections ("Software Updates") to the Software. If Sun, at its sole option, supplies Software Updates to you, the Software Updates will be considered part of Software, and subject to the terms of this Agreement.

5. NO SUPPLEMENTAL TERMS. The Supplemental Terms following the BCL do not apply to the Evaluation Agreement.

Sun Microsystems, Inc. Binary Code License Agreement

READ THE TERMS OF THIS AGREEMENT AND ANY PROVIDED SUPPLEMENTAL LICENSE TERMS (COLLECTIVELY "AGREEMENT") CAREFULLY BEFORE OPENING THE SOFTWARE MEDIA PACKAGE. BY OPENING THE SOFTWARE MEDIA PACKAGE, YOU AGREE TO THE TERMS OF THIS AGREEMENT. IF YOU ARE ACCESSING THE SOFTWARE ELECTRONICALLY, INDICATE YOUR ACCEPTANCE OF THESE TERMS BY SELECTING THE "ACCEPT" BUTTON AT THE END OF THIS AGREEMENT. IF YOU DO NOT AGREE TO ALL THESE TERMS, PROMPTLY RETURN THE UNUSED SOFTWARE TO YOUR PLACE OF PURCHASE FOR A REFUND OR, IF THE SOFTWARE IS ACCESSED ELECTRONICALLY, SELECT THE "DECLINE" BUTTON AT THE END OF THIS AGREEMENT.

1. LICENSE TO USE. Sun grants you a non-exclusive and non-transferable license for the internal use only of the accompanying software and documentation and any error corrections provided by Sun (collectively "Software"), by the number of users and the class of computer hardware for which the corresponding fee has been paid.

2. RESTRICTIONS. Software is confidential and copyrighted. Title to Software and all associated intellectual property rights is retained by Sun and/or its licensors. Except as specifically authorized in any Supplemental License Terms, you may not make copies of Software, other than a single copy of Software for archival purposes. Unless enforcement is prohibited by applicable law, you may not modify, decompile, or reverse engineer Software. You acknowledge that Software is not designed, licensed or intended for use in the design, construction, operation or maintenance of any nuclear facility. Sun disclaims any express or implied warranty of fitness for such uses. No right, title or interest in or to any trademark, service mark, logo or trade name of Sun or its licensors is granted under this Agreement.

3. LIMITED WARRANTY. Sun warrants to you that for a period of ninety (90) days from the date of purchase, as evidenced by a copy of the receipt, the media on which Software is furnished (if any) will be free of defects in materials and workmanship under normal use. Except for the foregoing, Software is provided "AS IS". Your exclusive remedy and Sun's entire liability under this limited warranty will be at Sun's option to replace Software media or refund the fee paid for Software.

4. DISCLAIMER OF WARRANTY. UNLESS SPECIFIED IN THIS AGREEMENT, ALL EXPRESS OR IMPLIED CONDITIONS, REPRESENTATIONS AND WARRANTIES, INCLUDING ANY IMPLIED WARRANTY OF MERCHANTABILITY, FITNESS FOR A PARTICULAR PURPOSE OR NON-INFRINGEMENT ARE DISCLAIMED, EXCEPT TO THE EXTENT THAT THESE DISCLAIMERS ARE HELD TO BE LEGALLY INVALID.

5. LIMITATION OF LIABILITY. TO THE EXTENT NOT PROHIBITED BY LAW, IN NO EVENT WILL SUN OR ITS LICENSORS BE LIABLE FOR ANY LOST REVENUE, PROFIT OR DATA, OR FOR SPECIAL, INDIRECT, CONSEQUENTIAL, INCIDENTAL OR PUNITIVE DAMAGES, HOWEVER CAUSED REGARDLESS OF THE THEORY OF LIABILITY, ARISING OUT OF OR RELATED TO THE USE OF OR INABILITY TO USE SOFTWARE, EVEN IF SUN HAS BEEN ADVISED OF THE POSSIBILITY OF SUCH DAMAGES. In no event will Sun's liability to you, whether in contract, tort (including negligence), or otherwise, exceed the amount paid by you for Software under this Agreement. The foregoing limitations will apply even if the above stated warranty fails of its essential purpose.

6. Termination. This Agreement is effective until terminated. You may terminate this Agreement at any time by destroying all copies of Software. This Agreement will terminate immediately without notice from Sun if you fail to comply with any provision of this Agreement. Upon termination, you must destroy all copies of Software.

7. Export Regulations. All Software and technical data delivered under this Agreement are subject to US export control laws and may be subject to export or import regulations in other countries. You agree to comply strictly with all such laws and regulations and acknowledge that you have the responsibility to obtain such licenses to export, re-export, or import as may be required after delivery to you.

8. U.S. Government Restricted Rights. If Software is being acquired by or on behalf of the U.S. Government or by a U.S. Government prime contractor or subcontractor (at any tier), then the Government's rights in Software and accompanying documentation will be only as set forth in this Agreement; this is in accordance with 48 CFR 227.7201 through 227.7202-4 (for Department of Defense (DOD) acquisitions) and with 48 CFR 2.101 and 12.212 (for non-DOD acquisitions).

9. Governing Law. Any action related to this Agreement will be governed by California law and controlling U.S. federal law. No choice of law rules of any jurisdiction will apply.

10. Severability. If any provision of this Agreement is held to be unenforceable, this Agreement will remain in effect with the provision omitted, unless omission would frustrate the intent of the parties, in which case this Agreement will immediately terminate.

11. Integration. This Agreement is the entire agreement between you and Sun relating to its subject matter. It supersedes all prior or contemporaneous oral or written communications, proposals, representations and warranties and prevails over any conflicting or additional terms of any quote, order, acknowledgment, or other communication between the parties relating to its subject matter during the term of this Agreement. No modification of this Agreement will be binding, unless in writing and signed by an authorized representative of each party.

FORTE(TM) FOR JAVA(TM), RELEASE 3.0, ENTERPRISE EDITION SUPPLEMENTAL LICENSE TERMS

These supplemental license terms ("Supplemental Terms") add to or modify the terms of the Binary Code License Agreement (collectively, the "Agreement"). Capitalized terms not defined in these Supplemental Terms shall have the same meanings ascribed to them in the Agreement. These Supplemental Terms shall supersede any inconsistent or conflicting terms in the Agreement, or in any license contained within the Software.

1. Software Internal Use and Development License Grant. Subject to the terms and conditions of this Agreement, including, but not limited to Section 3 (Java(TM) Technology Restrictions) of these Supplemental Terms, Sun grants you a non-exclusive, non-transferable, limited license to use internally the binary form of the Software complete and unmodified for the sole purpose of designing, developing and testing your Java applets and applications intended to run on the Java platform.

2. License to Distribute Redistributables. Subject to the terms and conditions of this Agreement, including but not limited to Section 3 (Java Technology Restrictions) of these Supplemental Terms, Sun grants you a non-exclusive, non-transferable, limited license to reproduce and distribute the binary form of those files specifically identified as redistributable in the Software "RELEASE NOTES" file ("Redistributables") provided that: (i) you distribute the Redistributables complete and unmodified (unless otherwise specified in the applicable RELEASE NOTES file), and only bundled as part of Programs, (ii) you do not distribute additional software intended to supersede any component(s) of the Redistributables, (iii) you do not remove or alter any proprietary legends or notices contained in or on the Redistributables, (iv) for a particular version of the Java platform, any executable output generated by a compiler that is contained in the Software must (a) only be compiled from source code that conforms to the corresponding version of the OEM Java Language Specification; (b) be in the class file format defined by the corresponding version of the OEM Java Virtual Machine Specification; and (c) execute properly on a reference runtime, as specified by Sun, associated with such version of the Java platform, (v) you only distribute the Redistributables pursuant to a license agreement that protects Sun's interests consistent with the terms contained in the Agreement, and (v) you agree to defend and indemnify Sun and its licensors from and against any damages, costs, liabilities, settlement amounts and/or expenses (including attorneys' fees) incurred in connection with any claim, lawsuit or action by any third party that arises or results from the use or distribution of any and all Programs and/or Software.

3. Java Technology Restrictions. You may not modify the Java Platform Interface ("JPI", identified as classes contained within the "java" package or any subpackages of the "java" package), by creating additional classes within the JPI or otherwise causing the addition to or modification of the classes in the JPI. In the event that you create an additional class and associated API(s) which (i) extends the functionality of the Java platform, and (ii) is exposed to third party software developers for the purpose of developing additional software which invokes such additional API, you must promptly publish broadly an accurate specification for such API for free use by all developers. You may not create, or authorize your licensees to create, additional classes, interfaces, or subpackages that are in any way identified as "java", "javax", "sun" or similar convention as specified by Sun in any naming convention designation.

4. Java Runtime Availability. Refer to the appropriate version of the Java Runtime Environment binary code license (currently located at http://www.java.sun.com/jdk/index.html) for the availability of runtime code which may be distributed with Java applets and applications.

5. Trademarks and Logos. You acknowledge and agree as between you and Sun that Sun owns the SUN, SOLARIS, JAVA, JINI, FORTE, and iPLANET trademarks and all SUN, SOLARIS, JAVA, JINI, FORTE, and iPLANET-related trademarks, service marks, logos and other brand designations ("Sun Marks"), and you agree to comply with the Sun Trademark and Logo Usage Requirements currently located at http://www.sun.com/policies/trademarks. Any use you make of the Sun Marks inures to Sun's benefit.

6. Source Code. Software may contain source code that is provided solely for reference purposes pursuant to the terms of this Agreement. Source code may not be redistributed unless expressly provided for in this Agreement.

7. Termination for Infringement. Either party may terminate this Agreement immediately should any Software become, or in either party's opinion be likely to become, the subject of a claim of infringement of any intellectual property right.

For inquiries please contact: Sun Microsystems, Inc. 901 San Antonio Road, Palo Alto, California 94303 (LFI#91206/Form ID#011801)